placer examination

PRINCIPLES
AND
PRACTICE

by
John H. Wells
Mining Engineer
Bureau of Land Management

ACKNOWLEDGEMENTS

The author is indebted to the Bureau of Land Mangement for making the publication of this book possible.

A book of this type would not be complete without the illustrated description of placer drilling and related data contained in Appendix D. These were provided by Mr. Cy Ostrom of C. Kirk Hillman Company, whose generosity is appreciated.

CONTENTS

PREFACE

As the title suggests, the subject matter in this book deals for the most part with the examination and testing of placers. It is intended primarily as a guide for the professional mineral examiner who examines mining claims located on public lands of the United States.

The book has been divided into eight parts or sections. Parts I and II, which deal with placer-forming processes and types of placers, are intended as a review of placer fundamentals. Other sections bring together the fundamentals of placer investigation and industry-proven testing procedures.

Part VIII is a glossary of the placer industry, setting out more than three hundred placer-related terms and definitions, many of which have not previously appeared in glossary form.

In scope, this book covers problems of placer investigation from initial reconnaissance to computing the value of a blocked-out placer. It is intended to serve as a reference book as well as a manual for day-to-day work. Its use should help the mineral examiner identify his sampling problems and select procedures appropriate to the purpose of his examination and to the deposits involved.

While this book has been prepared specifically for use by Bureau of Land Management personnel, it is hoped that it will at the same time be useful to the mining profession generally.

PART I

REVIEW OF PLACER THEORY
AND GEOLOGY

1. PLACERS DEFINED

In the United States mining law, mineral deposits that are not veins in place are treated as placers so far as locating, holding and patenting are concerned. But for the purposes of this book, the term "placer" is applied to deposits of sand, gravel, and other detrital or residual material containing a valuable mineral which has accumulated through weathering and mechanical concentration processes. Prerequisites for a placer are:

A valuable mineral which is relatively heavy and is resistant to weathering and abrasion.

Release of the valuable mineral from its parent rock.

Concentration of the valuable mineral into workable deposits. This step usually involves water transport.

The term "placer" as used in this book applies to ancient (Tertiary) gravels as well as to recent deposits, and to underground (drift mines) as well as to surface deposits.

2. STUDY OF PLACERS - GENERAL

The study of placers is not simple—it involves many of the disciplines of geology, with special emphasis on the theory and habits of streams. Although the location, size, and shape of a placer will reflect the regional forces of erosion, transportation, and deposition which created it, its final form will be controlled or modified by purely local conditions. As a result, each placer deposit can be expected to be unique in one or more ways and the field investigator should approach his work with this in mind.

All placers begin with weathering and disintegration of lodes or rocks containing one or more heavy, resistant minerals such as gold, platinum, magnetite, garnet, zircon, cassiterite, monazite, etc. It should be stressed that the end richness and size of a placer deposit will depend more on there being an abundant supply of source materials, and on conditions favorable for their concentration, than on the actual richness of the primary source. While the following paragraphs refer largely to gold placers, the principles set forth apply to all types. At the risk of oversimplification, a brief discussion of basic placer-forming processes will be taken up under the following headings:

Sources of valuable mineral.
Weathering and release processes.
Stream processes related to placers.
Concentration of valuable minerals.
Preservation of the deposit.

3. SOURCES OF VALUABLE MINERAL

The source of gold or other minerals found in a placer may be one or more of the following:

    a. Lodes or mineralized zones.

    b. Erosion of pre-existing placer deposits.

    c. Low-grade auriferous conglomerates or glacial debris.

    d. Magmatic segregations and associated basic rocks.

    e. Regional rocks containing scattered particles of valuable mineral.

    a. *Lodes*: Although placers are commonly found in lode mining districts, experience has shown that there is no fixed relation between the richness of the parent lode and the richness or size of resultant placers. Some of the most noted gold mining districts such as Goldfield, Nev., contain no significant placers. On the other hand, some highly productive placer areas are not associated with valuable lodes, the Klondike region being an example. In some cases the lode source may have been completely removed by erosion, while in others, it can be demonstrated that the gold or other valuable mineral was not derived from a single source, but instead, from many small mineralized seams or zones scattered through the bedrock. Although individually unimportant, these smaller sources can collectively furnish substantial amounts of gold.

    b. *Pre-existing placers*: In many localities old placers have been destroyed by erosion and their gold content reconcentrated in present day streams. Excellent examples are found in the Sierra Nevada region of California. Here, in Tertiary times, beginning some 60 million years ago, there existed an extensive river system, a prolonged time of rock decay, and a good balance between erosion and deposition. These factors combined to form widespread gold placers and some extremely rich bedrock concentrations in the Tertiary streams. Later, as the present Sierra Nevada range was elevated by mountain-building disturbances, new streams flowing toward the west cut across the ancient channels and swept some of their gold concentrations into today's streams where new, and sometimes richer, concentrations accumulated. Somewhat similar reconcentrations were found in parts of Alaska, the Klondike, and other places.

    c. *Auriferous conglomerates and glacial debris*: In some placer areas the immediate source of gold is found to be a low-grade auriferous conglomerate or a glacial deposit. With few exceptions such material is too poor to be worked but where appreciable quantities have been eroded, superficial placers may form on the erosion surface or within the local drainage. Wind erosion of an auriferous conglomerate occasionally creates an enriched surface veneer amenable to small-scale, selective mining. Under favorable conditions, gold derived from low-value glacial debris is carried into streams and reconcentrated into minable placers and although some of these reconcentrations have an appreciable extent, most are relatively small and spotty.

Examples of placers derived from low-grade auriferous conglomerates are found in the Red Rock, Goler, and Summit mining districts of Kern County, Calif., (Hulin, 1934) [1]. At Breckenridge and at Fairplay, Colo., gold derived from glacial moraines has been successfully dredged along the Swan, Blue, and South Platte Rivers. According to MacDonald (1910), a gold-bearing glacial till in the Weaverville district of Trinity County, Calif., is an important source of gold for the Trinity River and its tributaries.

d. *Magmatic segregations*: Disseminated deposits associated with magmatic segregations in basic rocks seldom generate significant placers but a notable exception is found at Goodnews Bay, Alaska, where extensive platinum placers have been derived from a large mass of dunite and related basic rocks. Although minable placers from basic or ultrabasic rocks may be few in number, they can be important, as shown by the Goodnews Bay platinum deposits where dredging has been carried on since 1937.

e. *Regional rocks*: Monazite placers of the type dredged in central Idaho, provide an excellent example of a placer mineral originating as small, sparsely distributed particles in the country rock. Here, monazite, which is a rare-earth mineral chiefly valuable for its thorium content, originates as an accessory mineral in the regional granitic rocks. Following liberation by the usual weathering processes it finds its way into stream deposits, where because of its moderately high specific gravity, it accumulates in placers along with other heavy minerals. Similarly, most of the magnetite, ilmenite, rutile, garnet and zircon associated with gold placers, could be traced to the regional bedrock where such minerals commonly occur as scattered particles or accessory minerals.

4. WEATHERING AND RELEASE PROCESSES

The first step in placer formation is release of valuable mineral from its parent rock. The various phenomena which combine to decompose and disintegrate rocks are embraced in the general term "weathering."

The chief agents of rock weathering are:

a. Ground water.
b. Temperature change.
c. Plant growth.
d. Surface erosion.

a. *Ground water*: The great solvent of rocks is, of course, water; particularly ground water that has become charged with products of

---

[1] References cited will be found at the end of Part I.

mineral and vegetable decay. Some rocks such as limestone may be completely dissolved but the majority are more commonly decomposed through the processes of oxidation, carbonation, and hydration.

b. *Temperature*: Change of temperature combined with frost wedging plays a part in rock weathering, particularly in desert regions where changes may be rapid and extreme. The effects of frost wedging are readily apparent where it loosens large blocks, but its lesser-known and more important function is the wedging apart of minute rock scales and mineral grains, which in some cases, is an effective agent of rock destruction.

c. *Plant growth*: Plants contribute to rock destruction and mineral release in two ways.
   (1) Root growth forces open and extends surface and near-surface cracks and crevices.
   (2) Decaying plants contribute rock solvents to the percolating ground waters. Decomposing vegetation is a very powerful agent of rock decay and, consequently, wet, humid climates favor the development of deep mantles of decomposed rock and the liberation of placer-forming minerals.

d. *Surface erosion*: On the steeper slopes surface erosion becomes more effective and, in turn, chemical rock decay assumes a lesser role in the weathering process. The rapid surface erosion of relatively unweathered rocks in desert regions can result in the formation of placers, but as shown by Lindgren (1933, p. 219), such placers are not often rich or highly concentrated.

5. **STREAM PROCESSES RELATED TO PLACERS**

Because streams are the dominant agent in the formation of most placer deposits, the field engineer should understand the fundamental processes involved in the movement and deposition of sediments by flowing water. In particular, he should be able to relate stream processes to the mechanics of placer formation. Actually this is a complex subject but several of the more important concepts of stream processes related to placer formation are:

a. Stream bed erosion occurs only when the flow has the ability to transport more material than is being supplied, and conversely, deposition will begin when the stream becomes overloaded or has its velocity sufficiently checked. Temporary or semi-permanent deposits are usually formed where velocity is locally reduced on the downstream side of rocks or other channel obstructions and along the inside banks of curves. Although at any one point in a stream the balance may be precarious, alternating between erosion and deposition, the net effect will be one or the other. Where there is a net reduction the stream is said to be degrading, and where there is a net deposition it is said to be aggrading.

b. It can be stated generally that streams do most of their transporting while in flood, and most of their depositing when the floods recede. In general, the coarse material transported by a stream moves intermittently and much of the coarsest material may be at rest for all but brief periods of time. Other things being equal, the length of time between moves and the distance moved is largely a function of particle size.

c. Stream conditions wherein the entire bed load is shifted or is agitated sufficiently to allow a downward movement of gold particles may occur only during extreme flood conditions or periods of regional climatic change, many years or centuries apart. Nevertheless, bed scour or movement of the whole contents of a stream bed, or of all except the large boulders, is essential to formation of the typical stream placer. Concentration of gold on the bedrock would not be possible without channel scour or a general agitation of the bed load.

d. Today's climate and present stream conditions may have little or no resemblance to those which controlled formation of a given placer. Also, it should be borne in mind that old stream deposits such as those of Tertiary age may have no relation to today's topography or drainage pattern.

e. In the case of desert placers, the usual rules of stream deposition and of placer concentration are often subordinate to other factors. These conditions are discussed under the heading of DESERT PLACERS.

Longwell, Knopf and Flint (1948) present in concise language, a broad picture of fluvial processes and the movement of alluvial material by streams. Although this reference does not discuss the formation of placers it can serve as a quick review of fundamental principles. Jenkins (1946), in an article titled "New Technique Applicable to the Study of Placers," discussed stream processes and their relation to placers in considerable detail. This well illustrated article is possibly the best single reference on the subject.

6. CONCENTRATION OF VALUABLE MINERALS

In theory, the richest part of a placer should be found near bedrock and, because of this, many people think that gold placers invariably are richer on bedrock than elsewhere within the deposit. On this basis, they believe that if indications of value are found in the upper horizons, pay gravel will surely be found on bedrock and countless mining ventures have been launched on this premise. Needless to say, many failed. In practice, it is not uncommon to find deposits in which the pay is scattered through a gravel mass without a significant bedrock enrichment. Some deposits, in fact, have their concentrations at the surface rather than on bedrock. This type of occurrence is explained in the section titled "Flood Gold Deposits."

7

Another popular idea is that concentration of gold in a stream is analogous to concentration in a sluice box. But in nature the process is by no means this simple as will be shown in following paragraphs.

a. *Bedrock concentration*: The development of bedrock concentrations in simple stream deposits may be generalized as follows: Consider a down-cutting stream in which the bed load of sand, gravel and boulders is progressively shifting downstream and subjected to general agitation during times of high water. During high-water periods some of the smaller and lighter particles will be picked up and advanced by the current, usually in a series of short jumps, while the heavier rocks and boulders roll and slide along the bottom. Each time the bed load moves or is generally agitated, the contained gold particles can work downward toward bedrock. Under favorable conditions this may be quite rapid but in many deposits, such bed movements occur only at long-spaced intervals—perhaps not within the lifetime of any one observer. Nevertheless, it must be realized that without such movement and rearrangement of the entire bed load, a downward migration of the gold and, in particular, its eventual concentration on bedrock would not be possible. Under flood conditions fine gold may be swept up and carried away but once coarse gold has settled to the bedrock it is very difficult for the current to dislodge it. For this reason coarse gold is generally found near its source.

b. *Types of bedrock*: The ultimate richness of a placer is dependent to a large extent on the physical characteristics of the bedrock. Slates and jointed rocks, particularly those dipping at steep angles are considered most effective in capturing and holding gold. Compact clay, clayey volcanic tuff and decomposed granite make effective bedrocks. A smooth, hard bedrock such as serpentine is generally considered to be a poor gold saver. In some important placer fields, the pay gravels rest on a false bedrock. In the Folsom, California, area, for example, this is usually a layer of volcanic tuff found well above the true bedrock. Gold works its way into soft or decomposed bedrocks and settles into the cracks and crevices of hard formations. It commonly migrates downward to an extent requiring the mining of several feet of bedrock to effect a complete gold recovery. In some cases, inability to dig a hard, rough bedrock has been the direct cause for failure of dredging ventures. In others, having to dig unexpected amounts of soft bedrock has seriously upset the initial cost estimates. Such possibilities should always be considered when evaluating a placer property.

c. *Pay Streaks*: Bedrock enrichments known as "pay streaks" usually follow a sinuous course and sometimes have no discernible relation to the present stream channel. A pay streak may split or terminate suddenly and its lateral limits may be irregular or indefinite, its location and eccentricities being dictated by local controls. Bear in mind that conditions which caused the pay streak, particularly in the case of older deposits, may no longer be in evidence. Pay streaks in gravel-plain or similar widespread-type placers are usually less definite than those in stream placers.

7. **PRESERVATION OF THE DEPOSIT**

Placer deposits do not suddenly come into being but, instead, they are progressively accumulated and enriched by a succession of stream actions and interactions. Although the land forms and geologic controls responsible for the creation of a particular placer may persist for a long period of time (as measured by the affairs of man), the deposit itself is transitory. Unless preserved by some disruption of the normal erosion cycle, the very forces which create the placer will in time destroy it. Knowing how and why a particular deposit was preserved is important to the prospector or mine operator and, obviously, such knowledge is also important to the examining engineer—particularly when he is called upon to project the course of a buried channel or appraise its potential. The more common causes of preservation are:

a. *Abandonment*: This is probably the most frequently encountered form of preservation. It is brought about where a stream, by change of course or rapid downward cutting, abandons its former bed and the placers therein. The abandoned placers are preserved in the sense that they are no longer subject to erosion by the parent stream. High bench and terrace gravels provide typical examples of abandoned placers.

b. *Regional uplift*: Elevation of the ocean shoreline has in some places preserved beach placers by raising them above the active surf. In some localities, such as Nome, Alaska, elevated beach placers are now several miles inland.

c. *Burial*: The lava-capped Tertiary channels found in the Sierra Nevada region of California offer some of the best examples of buried placers but there are means other than volcanic flows by which burial and preservation may be affected. Among these are:

Covering by landslide material.
Covering by flood-plain gravels following regional subsidence or a net change in climate.
Burial under alluvial fans or outwash detritus in desert regions.

Needless to say, a knowledge of past geologic or physiographic events is an important tool in the study of placers.

REFERENCES CITED (PART I)

Hulin, Carlton D., 1934, Geologic features of the dry placers of the northern Mojave Desert: Thirtieth Report of the State Mineralogist, pp. 417-642, California Division of Mines.

Jenkins, Olaf P., 1946, New technique applicable to the study of placers: California Division of Mines Bull. 135, Placer mining for gold in California, pp.149-216.

First published in California Journal of Mines and Geology, Vol. 31, No. 2, 1935 pp. 143-210.

Reprinted as a series of 8 articles in Mineral Information Service, Vol. 17, Nos. 1, 2, 3, 4, 5, 6, 8 and 9, Jan.-Sept., 1964. California Division of Mines.

Lindgren, Waldemar, 1933, Mineral deposits, 4th ed: McGraw-Hill Book Co., New York.

Longwell, Chester R.; Knopf, Adolph and Flint, Richard F., 1948, Physical geology, 3rd ed: John Wiley & Sons, New York.

MacDonald, D.F., 1910, The Weaverville-Trinity Center gold gravels, Trinity County, California: U.S. Geol. Survey Bull. 430, pp. 48-58.

PART II

TYPES OF PLACERS

Perhaps the best known schemes for classifying placer deposits are those by Jenkins (1946, p. 161) and Brooks (1913, pp. 25-32) 1/, the former being based on conditions of deposition and the latter on present position of the deposit. The field engineer should acquaint himself with these schemes, particularly that by Jenkins which has been developed in some detail. For the usual field investigation a somewhat simpler classification will be found suitable. This is:

1. Residual placers.
2. Eluvial placers.
3. Stream placers.
4. Bench placers.
5. Flood gold deposits.
6. Desert placers.
7. Tertiary gravels.
8. Miscellaneous types.
   a) beach placers.
   b) glacial deposits.
   c) eolian placers.

1. RESIDUAL PLACERS

A residual placer is, in effect, a concentration of gold (or other heavy mineral) at or near its point of release from the parent rock. In this type of placer the enrichment results from the elimination of valueless material rather than from concentration of values brought in from an outside source. Residual placers may be rich but they are not likely to be large and as a class, they have been relatively unimportant. The "seam diggings" in El Dorado County, Calif., (Clark and Carlson, 1956, p. 435) offer an example of residual gold placers.

2. ELUVIAL PLACERS

Eluvial placers usually represent a transitional stage between a residual placer and a stream placer. Where one type merges into another, they cannot be clearly distinguished. They are characteristically found in the form of irregular sheets of surface detritus and soil mantling a hillside below a vein or other source of valuable mineral. It should be noted that the parent vein or lode may or may not outcrop at the actual ground surface. Eluvial placers differ from residual placers in that surface creep slowly moves the gold and weathered detritus down hill, allowing the lighter portions to be removed by rain wash and wind. As the detrital mass gravitates downhill, a rough

---

1/ References cited will be found at the end of Part II.

stratification or concentration of values may develop but this is rarely perfected to the degree found in stream placers. Eluvial placers are typically limited in extent but there have been cases such as at Round Mountain, Nevada, (Vanderburg, 1936, pp. 133-145) where this type of placer supported large-scale mining operations.

3. **STREAM PLACERS**

Stream placers are the most widespread type in the Western States and, accordingly, are the type most frequently encountered in mineral examinations. Individual deposits vary so much that few general statements can be made concerning them but for the purpose of this review, they can be conveniently divided into:
   a. Gulch placers.
   b. Creek placers.
   c. River deposits.
   d. Gravel-plain deposits.

a. *Gulch placers*: Gulch placers are characteristically small in area, have steep gradients and are usually confined to minor drainages in which a permanent stream may or may not exist. This type of placer is, as a rule, made up of a mixture of poorly sorted gravel and detritus from adjacent hillsides. Because of steep gradient, the gravel accumulations are often thin and discontinuous. Boulders are commonly found in quantities that preclude all but simple hand mining operations. The gold is likely to be coarse and well-concentrated on bedrock. Gulch placers were usually the first to be found by the early miners and because most can be worked with simple hand tools, unworked remnants of shallow gulch deposits are not likely to contain material that would yield a profit today. The early-day miner was generally well-schooled by experience, and a dilligent worker. Any pay gravel that he left was usually cleaned up by the patient Chinese who followed. This was particularly true in gulch placers.

b. *Creek placers*: In many districts creek placers have been important sources of gold but like the gulch placers most were carefully prospected by the early miners and worked out, where worthwhile to do so. Many of the lower-grade remnants left by the early hand miners have since been exploited by some form of mechanized mining, notably by dragline dredges during the depression years of the 1930's. Creek placers as a group no longer contain significant economic reserves in the Western States but some in Alaska are mined with nonfloating washing plants and moveable sluices utilizing various combinations of hydraulic and mechanical excavation equipment.

c. *River deposits*: River deposits are represented by the more extensive gravel flats in or adjacent to the beds of present-day rivers and as a class, they have been our most important source of placer minerals. They are generally similar to creek placers but the gold is usually finer, the gravel well-rounded and large boulders fewer or absent. Although the over-all deposit may be low-grade, pay streaks and bedrock concentrations capable of supporting dredging or other large-scale mining operations are not uncommon. At many places in California, the early miners diverted rivers through tunnels or bypassed the water in flumes to permit mining the river bed. In this manner, many miles of the middle and upper reaches of the principle gold-bearing rivers were effectively cleaned out. The lower reaches of many of these streams were systematically dredged and, at one time, where conditions were favorable, gravels returning five cents per cubic yard were dredged at a profit. Needless to say, few important river deposits remain unknown in the United States.

d. *Gravel-plain deposits*: These are somewhat difficult to define as they may grade from river or bench deposits, into flood-plain or delta-type deposits and they can be geologically old, or recent. Gravel plains are found where a river canyon flattens and widens or, more often, where it enters a wide, low-gradient valley. The contained placers are generally similar to those in river deposits except for greater size and a more general distribution of gold. Because gravel-plain deposits are built by shifting stream channels, their gold is apt to have a wide lateral and vertical distribution and because of the relatively low velocities of streams flowing over flood plains, their placers are commonly made up of smaller-size gold compared with that found in the main stream deposits. Any larger gold carried by the main channel will likely be dropped close to the upper edge of the flood plain where the stream's velocity decreases and its transporting ability is reduced. Although subject to surface wash and flood erosion, most gravel-plain deposits are relatively permanent. Examples of this type of deposit are the dredging fields at Hammonton, and those near Folsom, California. Each produced gold valued at approximately $100,000,000.

4. BENCH PLACERS

Bench placers are usually remnants of deposits formed during an earlier stage of stream development and left behind as the stream cuts downward. The abandoned segments, particularly those on the hillsides, are commonly referred to as "bench" gravels. Frequently there are two or more sets of benches in which case the miners refer to them as "high" benches and "low" benches. In California and elsewhere, most bench deposits were quickly found by the early miners who proceeded to work the richer bedrock streaks by primitive forms of underground mining. At the time these were referred to as "hill diggings." Following the development of hydraulic mining in the 1850's, many of the larger bench deposits were worked by hydraulicking and the smaller ones by ground sluicing. During the depression years of the 1930's, much of the so-called "sniper" mining was carried out on remnants of bench gravels and it should be noted that these hard-working individuals seldom recovered more than 25 or 30 cents per day.

5. FLOOD GOLD DEPOSITS

As a rule, finely-divided gold travels long distances under flood conditions. This gold which can best be referred to by the miners' term of "flood gold", consists mostly of minute particles so small that it may take 1,000 to 5,000 colors to be worth 1 cent. With few exceptions such gold has proven economically unimportant. The mineral examiner should recognize the true nature of flood gold deposits so that he can guard against being misled by their seemingly-rich surface concentrations.

As a stream sweeps around a curve, the water is subject to tangential forces which cause a relative increase in velocity along the outer radius of the curve with a corresponsing decrease along the inside radius. The bottom layer of water is retarded by friction and as a result, it has a tendency to flow sideways along the bottom toward the inner bank. This, in turn, causes sand and small gravel to accumulate in the form of an accretion bar along the inside bank of the curve and where flood-borne particles of gold are being carried down the stream, some will be deposited near the upper point of such bars, as shown in Figure 1.

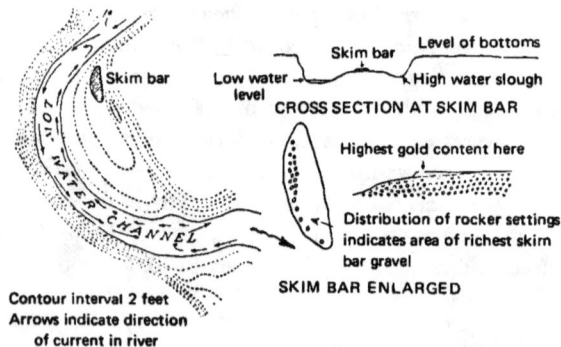

FIGURE 1. — Sketch showing the location of flood gold on accretion, or skim bars. (From USGS Bull. 620-L)

The foregoing is an oversimplification of a complex stream process but the fact is, in streams draining a gold-bearing region, seemingly rich deposits of fine-size gold may be concentrated near the upper point of the inside bars, between the high and low water marks. Good surface showings of fine-size gold are not uncommon and although they may appear to be valuable, experience has shown that in most cases the gravel a few inches beneath these surface concentrations is nearly worthless.

In many gold-producing areas, particularly in parts of South America, river bars have been skimmed by natives year after year from time immemorial.

These are not permanently exhausted because floods deposit a new supply of gold and the renewal will continue indefinitely. These are called Charcas de Oro which literally translates "Gold Farms."

The best-known flood gold occurrences in the United States are found along a 400-mile stretch of the Snake River in western Wyoming and southern Idaho. Here they are called "skim bars" (Hill, 1916) and have been intermittently worked since about 1860. See Figure 1. They were first exploited by transient miners employing rockers and simple sluicing operations and, by cleaning up the richer spots, a few did fairly well. It was inevitable that some would proceed to install dredges or other large washing plants and launch ambitious mining schemes on the strength of surface showings. Needless to say, following the exhaustion of superficial pay streaks, most of these large-scale mining schemes proved unprofitable. The foregoing is pointed out because flood gold concentrations are still to be found and without doubt, new mining ventures will be proposed and attempted from time to time, particularly by advocates of suction dredges. The mineral examiner should learn to recognize flood gold deposits and equally important, he should be fully aware of their pitfalls.

6. **DESERT PLACERS**

Desert placers in the Southwest occur under widely varying conditions but taken as a whole, they are so different from normal stream placers as to deserve a special classification. When dealing with the usual desert placer the mineral examiner must learn to disregard some of the rules of stream deposition, or at least, he must learn to apply them with caution. Desert placers are found in arid regions where erosion and transportation of debris depends largely on fast-rising streams that rush down gullies and dry washes following summer cloudbursts. During intervening periods, varying amounts of sand, gravel or side-hill detritus is carried in from the sides by lighter, intermittent rain wash which is sufficient to move material into the washes but not carry it further. When the next heavy rain comes, a torrential flow may sweep up all of the accumulated detrital fill, or only part of it, depending on intensity and duration of the storm and depth of fill. It should be obvious that the intermittent flows provide scant opportunity for effective sorting of the gravels or concentration of gold. Under such condtions the movement and concentration of placer gold will be extremely erratic. Moreover, where the entire bedload is not moved, any gold concentration resulting from a sudden water flow will be found at the bottom of the temporary channel existing at that time. This may be well above bedrock.

Desert miners have learned from experience that gold enrichments are sometimes found resting on caliche layers, particularly those near the ground surface, but such surface or near-surface concentrations are commonly small, residual-type accumulations of gold left behind where lighter material has been removed by rain wash and wind action. In other words, such enrichments result from the removal of valueless material rather than from the concentration of gold by normal stream processes. It should be stressed that in some desert placers the only economically minable ground is related to superficial concentrations and, at best, the chance of finding pay gravel is to a great extent fortuitous and largely dependent on careful prospecting.

Descriptions of many desert placer areas in the Southwest can be found in a number of publications among which are those published by the Arizona Bureau of Mines (Wilson and Fansett, 1961), the University of Nevada (Vanderburg, 1936) and the California Division of Mines (Haley, 1923, pp. 154-160).

7. TERTIARY GRAVELS

Gold-bearing gravels of Tertiary age are abundant in the central Sierra Nevada region of California and to a lesser extent in northwestern California and southwestern Oregon. A few are found in northeastern Oregon and central Idaho. In California, many miles of these ancient river channels have been mined and according to figures offered by (Clark 1965, p. 44), seven of the California deposits have together produced gold amounting to more than 232 million dollars. The total production is not on record but it is known to far exceed this figure. Although few, if any, of the remaining Tertiary deposits can be mined at the present price of gold, the possibility of mining some at a future date is real and must be given serious consideration. Because of this future potential and today's sporadic but continuing mining efforts, it is important that the mineral examiner have a clear concept of this special type of placer. The subject of Tertiary channels is too extensive for review here and, for this reason, the reader is directed to the following selected references:

Tertiary Channels (Clark, 1965, pp. 39-44). This article contains a brief history of mining operations and descriptions of the Tertiary channels of the Sierra Nevada region. Contains indexed map with list of principal deposits and bibliography.

California's Gold-Bearing Tertiary Channels (Jenkins & Wright, 1934). A condensed but excellent discussion of the origin of California's Tertiary gravels and related geology. Highlights factors in gold accumulation and discusses methods of geologic exploration and proposed geophysical techniques as aids to future exploration work.

Tertiary Gravels of the Sierra Nevada of California (Lindgren, 1911). A classic on Tertiary gravels and related subjects. Most writers on placer subjects have drawn heavily from this source of information and as a result, quotations from and references to Lindgren's work are prominent in the literature.

The Auriferous Gravels of the Sierra Nevada of California (Whitney, 1880). Discusses in considerable detail, the concepts (circa 1879) of origin and deposition of Tertiary gravels. The book is best suited to advanced study. Also contains maps and excellent descriptions of important Sierra Nevada placer mining districts and offers some of the best available data on individual mining operations of the day. Includes chapters by W.A. Goodyear and W.H. Pettee.

The Ancient River Beds of the Forest Hill Divide (Browne, 1890). One of many excellent articles describing Tertiary gravel deposits and related mining operations that have been published by the California State Mining Bureau.

Tertiary Gold-Bearing Channel Gravel in Northern Nevada County, California (Peterson and others, 1968). A report on studies of the San Juan Ridge Tertiary channel in Nevada County, California, made as part of the Department of the Interior's Heavy Metals Program. It describes the geology and physical characteristics of a 15-mile section of this channel and also discusses geophysical investigations of the channel gravel carried out by personnel of the United States Geological Survey. Seismic, resistivity, gravity, magnetic, electromagnetic, and induced polarization geophysical methods were applied and evaluated. 22 pp., maps, illus.

Gold Resources in the Tertiary Gravels of California (Merwin, 1968). One of a series of reports presenting results of investigations conducted by the United States Bureau of Mines under the Department of the Interior's Heavy Metals Program. It reviews the history of hydraulic and drift mining in California's Tertiary gravels and the reasons for cessation of large-scale mining of these deposits. The report discusses general geology, the distribution of gold within the gravels, past production, and gold reserves. 14 pp., illus., bibliography.

Most of the foregoing references are of a technical nature and primarily of interest to engineers and geologists.

The term "Tertiary gravel" is used broadly to designate extensive gold-bearing gravel deposits laid down in ancient streams approximately 50 million years ago. These often-rich channels were subsequently buried under as much as 1500 feet of younger gravels and volcanic material which effectively sealed and preserved the original placers. The present Sierra Nevada range and its Tertiary deposits have since been elevated by mountain-building disturbances and, today, the dissected channels are typically found as lava-capped segments high above the present streams. Parts of the Tertiary deposits were worked extensively by the hydraulic method during the latter half of the 1800's and in many places, the richer bedrock gravels were exploited by tunneling operations known as "drift mines", some of which followed the ancient channels for miles along their subterranean course. Parts of California's Tertiary channel system remain buried and unexplored; these are believed to contain a large gold reserve, available for mining at some future date.

8. MISCELLANEOUS TYPES

There are other less publicized types of gold placers that are not economically important today, but which may achieve importance at some future date. For this reason a brief discussion and some selected references are offered here.

a. *Beach placers*: Beach placers may form where gold-bearing material is carried into the ocean by streams, or along the wave-cut base of a gold-bearing coastal plain. With exception of the highly productive beach placers discovered at the turn of the century at Nome, Alaska, none have been of great importance in the Western States. Typical beach placers along the Pacific coast are found as erratically distributed, somewhat lenticular concentrations or streaks of black sand minerals with varying amounts of finely-divided gold and in some places, with platinum-group minerals. Beach-placer black sands can be expected to consist largely of magnetite and ilmenite but significant amounts of chromite are found in some Oregon beach sands. In the case of gold-bearing beach placers, the individual black sand concentrations are seldom over 100 feet long or more than a few feet thick. Those found on the active beaches are the result of storm and tidal action, and they come and go with changing conditions of the beach. Some of the most productive placers have been found in ancient, elevated beaches that are now several miles inland.

Beach mining reached its height following discovery of the rich gold placers at Nome, Alaska, in 1898. Here, over two million dollars were produced from a 20-mile section of modern beach about 200 feet wide, and another 15 million from a series of inland elevated beaches. Subsequent discoveries on other Alaskan beaches, and elsewhere along the Pacific coast, resulted in other gold rushes but little production. Some beach placers along the California-Oregon coast have been worked in the past by simple hand mining methods and although a number of large-scale operations have been attempted, none, to this writer's knowledge, have been successful.

Over the years a number of attempts have been made to mine magnetite-rich beach sands for their iron content and during World War II some chromite was recovered from ancient beach deposits in Oregon. (Kauffman and Baber, 1956, pp. 12, 13).

When investigating a beach placer, all reports and sampling data relating to the property should be examined critically. First it should be ascertained if the samples represent a minable volume, of if they actually represent selected or scattered streaks of enriched material. Second, it should be ascertained whether or not the reported gold values were determined by fire assay. In the absence of information to the contrary, it must be assumed that the reported values were obtained by fire assay. Little or no credence can be given sample results, or the reliability of a report, based on fire assay returns. There are persistent stories that beach deposits contain gold and platinum in forms not amenable to recovery by conventional processes, in fact, many of the larger-scale attempts to work beach deposits have been based on some kind of "revolutionary" or secret process intended to recover this elusive gold. The fact that few such schemes have gone into commercial production and none have become sustained operations is significant.

A number of excellent technical articles relating to beach deposits in Oregon, Alaska and California are available, among which are those by Pardee (1934); Dasher, Fraas and Gabriel (1942); Twenhofel (1943); Thompson (1915); Thomas and Berryhill (1962); and by Jenkins (1946). Recent work by the United States Geological Survey (Clifton, 1967) points out the problems normally encountered when sampling and analyzing beach or marine deposits containing detrital gold.

b. *Glacial deposits*: The mineral examiner working in the Western States may seldom encounter a placer directly associated with glacial deposits but, on the other hand, it is not unusual for a miner to assert that a particular deposit, particularly if its origin is obscure, is a "glacier" placer. For this reason the engineer should know about glacial deposits as they relate to placers.

The fundamentals have been well set out by Blackwelder (1932) as follows:

"Since it is the habit of a glacier to scrape off loose debris and soil but not to sort it at all, ice is wholly ineffective as an agency of concentration for metals. Gold derived from the outcrops of small veins is thus mixed with large masses of barren earth. Attempts to mine gold in glacial moraines, where bits of rich but widely scattered float have been found, are for that reason foredoomed to failure.

"If a glacier advances down a valley which already contains gold-bearing river gravel, it is apt to gouge out the entire mass, mix it with much other debris and deposit it later as useless till. Under some circumstances, however, it merely slides over the gravel and buries it without distributing it.

"On the other hand, the streams born of glaciers or slowly consuming their moraines have the power to winnow the particles of rock and mineral matter according to size and heaviness. Such streams may form gold placer deposits in the well-known way by churning the load they carry and allowing the heavy minerals to sink to the bedrock. Placers may therefore be found in the deposits of glacial rivers if there are gold veins exposed in the glaciated area upstream. Nearly all the gravel which has been dredged for gold along the foothills of the Sierra Nevada was deposited by rivers derived in part from glaciers along the crest of the range, but most

21

of the gold was probably picked up in the lower courses of such rivers. Since glacial rivers choke themselves and build up their channels progressively, their deposits are likely to be thicker and not so well concentrated as those of the more normal graded rivers which are not associated with glaciers."

Where a glacier-related placer is encountered, the field engineer should, as an early step in his investigation, search out and study all available technical literature relating to the glacial history of the region. In particular, he should seek any reliable information on past mining of the deposit or similar deposits in the district, the object being to determine if significant gold concentrations are to be expected and, if so, under what conditions they are likely to be found.

Two districts having glacier-related placers that are well described in the technical literature are those near Breckenridge and Fairplay, in Colorado. At Fairplay (Singewald, 1950), the actual moraines were mined locally but the most extensive and productive placers were found in outwash aprons extending away from the true moraines. At Breckenridge (Ransome, 1911, pp. 175-181), bench gravels associated with Pleistocene glacial deposits were mined by ground sluicing and hydraulicking while younger gravels derived from glacial moraines have been extensively dredged along the Swan and the Blue Rivers. Both districts are well described in the literature referred to.

In general, glacial debris that has been scoured from highly mineralized bedrock areas may be expected to contain gold but it will probably have little or no economic value unless resorted by post-glacial streams.

c. *Eolian placers*: In desert regions the wind may act as an agent of concentration by blowing sand and the lighter rock particles away from a body of low-value material and leaving an enriched surface veneer containing gold or other heavy minerals in a somewhat concentrated state. There have been many cases where wind-caused surface enrichments supported the activities of itinerant miners using hand tools and simple dry washers.

Although commercial-grade eolian placers are not likely to be encountered by today's mineral examiner, he should be aware of their existence and should be alert to their misleading appearance. In other words, when taking near-surface samples from desert placers, he should guard against unintentional salting which could result from the inclusion of non-representative, wind-caused surface enrichments.

REFERENCES CITED (PART II)

Blackwelder, Eliot, 1932, Glacial and associated stream deposits of the Sierra Nevada: California Div. of Mines, Mining in California, State Mineralogist Rept. 28, pp. 309-310.

Brooks, Alfred H., 1913, The mineral deposits of Alaska: Mineral Resources of Alaska, U.S. Geol. Survey Bull. 592, 1914.

Browne, Ross E., 1890, The ancient river beds of the Forest Hill divide: State Mineralogist Rept. X, California State Mining Bureau, pp. 435-465.

Clark, William B., 1965, Tertiary Channels: Mineral Information Service Vol. 18, No. 3, March 1965. California Division of Mines.

Clark, William B. and Carlson, Denton W., 1956, Mines and mineral resources of El Dorado County, California: California Journal of Mines and Geology. Vol. 52, No. 4, October 1956. California Division of Mines.

Clifton, H. Edward; Hubert, Arthur, and Phillips, R. Lawrence, Marine Sediment Sample Preparation for Analysis For Low Concentrations of Fine Detrital Gold: U.S. Geol. Survey Circular 545, 1967. 11 pp.

Dasher, John; Fraas, Foster and Gabriel, Alton, 1942, Mineral dressing of Oregon beach sands. Concentration of chromite, zircon, garnet and ilmenite: U.S. Bureau of Mines Report of Investigations 3668, 1942, p. 19.

Haley, Charles S., 1923, Gold placers of California: Bulletin 92, California State Mining Bureau, p. 167.

Hill, J.M., 1916, Notes on the fine gold of Snake River, Idaho: U.S. Geol. Survey Bull. 620-L, pp. 271-294.

Jenkins, Olaf P., 1946, New technique applicable to the study of placers: California Division of Mines Bull. 135, Placer mining for gold in California, pp. 149-216.

First published in California Journal of Mines and Geology, Vol. 31, No. 2, 1935 pp. 143-210.

Reprinted as a series of 8 articles in Mineral Information Service, Vol. 17, Nos. 1, 2, 3, 4, 5, 6, 8 and 9, Jan.-Sept., 1964. California Division of Mines.

Jenkins, Olaf P. and Wright Quinby W., 1934, California's gold-bearing Tertiary channels: Engineering and Mining Journal, Vol. 135, No. 11, November, 1934, pp. 497-502.

Kauffman, A. J., Jr. and Baber, K. D., 1956, Potential of heavy-mineral-bearing alluvial deposits in the Pacific Northwest: U.S. Bureau of Mines Information Circular 7767, p. 36.

Lindgren, Waldemar, 1911, Tertiary gravels of the Sierra Nevada of California: U.S. Geol. Survey, Professional Paper 73, p. 226.

Merwin, Roland W., 1968, Gold resources in the Tertiary gravels of California: U.S. Bureau of Mines Technical Progress Report, Heavy Metals Program, 1968.

Pardee, J.T., 1934, Beach placers of the Oregon coast: U.S. Geol. Survey Circular 8, 1934, p. 41.

Peterson, Donald W., Yeend, Warren E., Oliver, Howard W., and Mattick, Robert E., 1968, Tertiary gold-bearing channel gravel in Northern Nevada County, California: U.S. Geol. Survey Circular 566, 1968.

Ransome, Frederick L., 1911, Geology and ore deposits of the Breckenridge district, Colorado: U.S. Geol. Survey Professional Paper 75, 1911, p. 187.

Singewald, Quentin D., 1950, Gold placers and their geologic environment in Northwestern Park County, Colorado: U.S. Geol. Survey Bull. 955-D, 1950, pp. 103-172.

Thomas, Bruce I. and Berryhill, Robert V., 1962, Reconnaissance studies of Alaska beach sands, Eastern Gulf of Alaska: U.S. Bureau of Mines Report of Investigations 5986, p. 40.

Thompson,, Arthur G., 1915, Yakataga beach placers: Engineering and Mining Journal, Vol. 99, No. 18, May 1, 1915, pp. 763-765.

Twenhofel, W.H., 1943, Origin of the black sands of the coast of Southwest Oregon: State of Oregon Dept. of Geology and Mineral Industries Bull. 24, 1943.

Vanderburg, William O., 1936, Placer Mining in Nevada: University of Nevada Bull, Vol. 30 No. 4, May 15, 1936, p. 180.

Whitney, J.D., 1880, The auriferous gravels of the Sierra Nevada of California: Contributions to American Geology, Vol. 1: Harvard University Press, Cambridge, Mass.

Wilson, Eldred D. and Fansett George R., 1961, Gold placers and placering in Arizona: Arizona Bureau of Mines Bull. 168, p. 124.

PART III

SAMPLING AND EVALUATION

If the reader will pause and study Figure 2, he can better visualize the more pertinent placer sampling problems, as they are taken up in the following paragraphs.

FIGURE 2. — *Typical placer material showing the wide variation in particle sizes to be dealt with in sampling.*

1. GENERAL CONSIDERATIONS

    a. *Problems*: Contrary to popular belief, representative placer samples are seldom easy to obtain and in almost all cases sample results need a large measure of interpretation. Some of the underlying reasons are:

        (1) Large particle sizes to be dealt with. A representative sample should contain all of the constituents of a deposit and in exactly the same proportion in which they occur in the parent mass. But look at Figure 2. This deposit is a mixture of fine sand, pebbles and boulders varying from a few tens of pounds to hundreds of pounds, and in this respect, is typical of many placers. A little study of Figure 2 will show why when dealing with such deposits, it is virtually impossible to take a small sample representative of the

whole mass, and why the evaluation of such a deposit can tax the ability of an expert.

(2) High unit-value of gold. When dealing with the typical placer sample containing a high-value mineral such as gold, any error in mineral content of the sample will be highly magnified in the end result. Consider that in a commercial placer the relative amount of gold (by volume) may be on the order of one part gold to a hundred-million parts of gravel. [1] To the mineral examiner this means that a single fly speck of gold in a pan of gravel is equivalent to say, 2 to 5 cents per cubic yard, depending on the exact size of the speck. The actual separation of small amounts of such gold from overwhelming quantities of sand and gravel seldom presents a serious problem; the real problem is to take a sample that is representative in the first place.

(3) Erratic distribution of values. Obtaining a satisfactory placer sample would be a comparatively simple undertaking if the valuable minerals were uniformly distributed throught the whole mass. In almost all placers, however, the heavy minerals and particularly the gold are more or less segregated. For example, in cases where economic values are confined to pay streaks, these are likely to occur as narrow, discontinuous accumulations with perhaps little or no value in between. Where coarse gold is present it can be expected to be even more erratic in distribution and it should be evident that under such conditions reliable valuation will depend on something more than taking a few small samples and an exercise in arithmetic.

In theory, these problems can be overcome by taking samples large enough to offset the eccentricities of a deposit, but to do so, would mean taking samples measured in tons rather than pounds and it is seldom possible to take such large samples in actual practice.

Some approach the problem by arguing that if enough small samples are taken, the highs will balance the lows and the end figure will represent the average value of the deposit even though no one sample is correct. This may be statistically true but here again, practical considerations seldom permit taking the number of samples needed to achieve this end.

b. *Industry practice*: How do the established placer mining companies evaluate their prospects? First, the prospecting is put in charge of an employee who has had wide experience with the type of deposit or mining project involved. Second, they recognize that any one sample is rarely representative of the overall deposit and by itself may mean little. Third, after studying the combined prospecting data and perhaps taking special check samples, experience-based adjustments known as

---

[1] Ground having a gold-gravel ratio of $1:100,000,000$ (by volume) is worth approximately 15 cents per cubic yard (@ \$35/oz.)

"correction factors" are applied to the initial sample data where needed and, finally, the calculated values are based on these "adjusted" data. Here, it should be emphasized that the successful placer companies place as much reliance on the experience and insight of their prospecting and management personnel as they do on the sample results. If there is a mathematical formula or a general rule which will replace experience-based judgement, the operating companies have not found it.

c. *Minerals other than gold*: Parenthetically, it should be noted that placers chiefly valuable for minerals such as monazite, rutile, cassiterite, ilmenite, etc., are generally easier to sample and evaluate than gold placers. There are several reasons. First, the valuable mineral makes up a larger part of the mass. Second, the mineral sought has a relatively low unit value which means that extraneous particles of such mineral (in a sample) may have little effect on the calculated value. Third, deposits containing such minerals are often made up of well sorted, small-size detrital materials such as beach sands and, in such cases, the mechanics of sampling may be simplified by permitting the use of augers or jet drills in lieu of churn drills. Speaking generally, the sampler has considerably more latitude when dealing with minerals of low unit-value than when dealing with gold and his mistakes are less likely to seriously affect the end results.

d. *Other factors to be considered*: Many things other than mineral content have a direct bearing on the commercial value of a placer deposit. The mineral value in itself may be of secondary concern where sampling or examination of the lands for example, show unfavorable bedrock conditions, excessive amounts of sticky clay, large boulders or other factors that would adversely affect a placer mining operation. A good sampling program will provide the information needed to evaluate adverse physical conditions, should they exist, but here again experience is needed to interpret the information and make correct decisions.

e. *In brief*: Placer samples yield limited information. Correct interpretation of this information depends upon the engineer's powers of deduction and experienced judgement, rather than on the rigid application of formulae.

2. SAMPLING GUIDES

For practical reasons, placer sampling incorporates a series of steps which may be carried out one at a time or in some cases are more or less combined. These may be defined as:

Reconnaissance of the lands.
Sampling.
Sample processing.
Data processing.
Evaluation of results.

The first step can be further divided into three parts which will be discussed here under the following headings:

    a. Pre-sampling reconnaissance.
    b. Choosing a sampling method.
    c. Number and size of samples.

a. *Reconnaissance*: A preliminary inspection of the lands should precede any placer sampling work. If the property is small this may take less than an hour but if it is large or if it contains significant mineral exposures, several days of pre-sampling reconnaissance may be in order. The importance of this preliminary work should be self-evident for, obviously, before the engineer can make intelligent sampling plans he must have some idea of the size, shape and the general characteristics of the deposit being investigated. Unfortunately, this first step appears short-changed in many sampling programs.

If aerial photographs of the lands under study are available, copies should be obtained before starting the reconnaissance. Photographs of the type available from the U.S. Geological Survey or Forest Service are often the most useful reconnaissance tool. This is particularly true when used in stereoscopic pairs. Aside from providing a quick, overall perspective of the lands, good aerial photographs reveal details of topography, drainage and other features not easily discernible on the ground.

A review of published material relating to the lands under study should be part of any reconnaissance because information thus obtained can be a great time saver, particularly where the regional or local geology has been worked out by qualified observers and reported in authoritative publications. On the other hand, too much reliance on published material can lead to erroneous prejudgements, particularly among younger engineers not in a position to weigh the statements and conclusions of others against their own experience.

The history of a mining district can play an important role in the selection and interpretation of placer samples and should be considered an element of the reconnaissance. Any old diggings on the lands to be sampled merit careful consideration, particularly those from which some production has been made. Where production records are available, they often serve as a useful guide in the selection and interpretation of new samples. Nearby mines should also be visited and examined where possible to do so.

Because the reconnaissance must be tailored to suit the job at hand, no fixed procedures can be set out here. The effort and time required for an adequate job will depend on the examining engineer's experience and perception.

b. *Choosing a sampling method*: What sampling method should be used? This question comes up early in most placer investigations but there is no easy answer. Because each deposit has unique characteristics there is no single "best" method of sampling and no procedure can be applied universally. In some cases, decisions will have to be made on a sample-by-sample basis but a good pre-sampling reconnaissance will usually indicate if the ground should be tested by means of pits, shafts, churn drilling, or other means. It is a paradox that the very things which are to be determined by sampling are often things which govern the type of sample to be taken in the first place. This is a "chicken-and-egg" type of situation in which we sometimes cannot tell which comes first. Something to keep in mind is the fact that today's placer sampling is expensive. Because of this, the method or sampling program adopted should be no more elaborate than needed to provide the amount of detail and the end accuracy required for a particular determination. For example, the prospector, whose main concern is to expose enough of a "showing" to interest a second party or mining company in his prospect, would be foolish to expend his own time and money attempting to carry out an elaborate sampling program; that is, sampling which any cautious mining company would do over again for their own satisfaction and protection. Very often the first placer sampling is of a cursory nature sufficient only to indicate if further interest in the property is warranted or to serve as a guide in future sampling or exploratory work.

Some methods to be considered are:
  Sampling existing exposures.
  Use of hand-dug excavations.
  Machine-dug shafts.
  Backhoe pits or trenches.
  Bulldozer trenches.
  Churn drill holes.
  Bulk samples.
  Grab samples.

The order in which these methods are presented has no bearing on their relative importance or applicability in a given situation. Their applications are described in the section titled "Sampling Methods."

c. *Number and size of samples*: When investigating placers, the problem of how many samples are needed and where they should be taken, and how large they should be, is often a perplexing one. Here there is no formula, rule of thumb, or pat answer to reliably guide the sampler; in fact, placer sampling procedures may vary not only with every property but with the purpose of the examination and, to some extent, with the type of mining contemplated.

How can the mineral examiner cope with this problem? First, he must know what he is dealing with. Is it a large, regular deposit in which the gold or other valuable mineral is distributed somewhat uniformly? Or, is it a boulder-strewn, stream-type deposit containing coarse, erratically distributed gold? It is a generally accepted principle that the smaller and more uniform the size of the gravel, and the more evenly distributed the mineralization, the fewer samples needed for an intelligent estimate of value. While we may accept this principle as fact, the degree to which such characteristics affect sample size and the number of samples required remains largely a matter of judgement. To put it another way, textbooks tell us that for ordinary gold ore, the size of the largest piece of rock in the sample determines the weight of sample needed. According to Woodbridge (1916, p. 57) [1], if the largest piece is one inch, the minimum sample weight should be 2,000 pounds. Where the largest piece is two inches, the minimum weight should be 10,000 pounds, etc. If this progressive scale were applied to placer gravels, the required weights for typical samples would be measured in tens of tons. This points out that sampling procedures based on pure theory are too unwieldy for placer application and, in turn, it shows why the placer sampler must to a large extent rely on his own judgement and good sense to determine what is an adequate-size sample in a given case.

Now, consider a single sample. While a single sample may provide much information about the material it actually penetrates, it rarely, if ever, provides sufficient information for the valuation of a deposit. Where a single sample is taken its assumed area of influence will reflect the insight or optimism of the viewer. But, how do you measure these? There are placer deposits of such a character that it is safe to project sample data several hundred feet and there are others where the gold distribution is so erratic that a ten-foot projection would be a dangerous assumption.

But now suppose we take a second, third and fourth—or an extended series of samples. Obviously as each sample is taken, its area of influence extends only part way to adjacent samples and as more samples are taken at closer and closer intervals, the areas of influence become progressively smaller and the combined sample results progressively more representative until, in theory, a point is reached where the combined average equals the true value of the whole. This is known as sampling to a uniform average.

Sampling to a uniform average is simple in concept but its objective can seldom be reached in practice. In other words, for practical reasons the desired number of samples can rarely be taken and in the end, the placer operator must either reject a property or go forward on the strength of a valuation which is not 100 percent reliable and, hence, he must always

---

1/ References cited are listed at the end of Part III.

accept some degree of risk. Realistically then, the minimum number of samples might be considered the number required to hold this risk within acceptable limits. A method for graphically examining this problem has been described by Herr (1916, pp. 261, 262).

As a practical example, consider an actual sampling program carried out by an experienced gold dredging company. In this case the ground had a history of superficial hand mining operations going back many years and the geologic and physical conditions indicated a possible depth and volume sufficient to support a bucket-line dredge. A comprehensive drilling program was needed to determine the average tenor and to define possible mining limits.

As a first step, a 6-inch placer drill was moved to the property and four holes drilled at random points to test the depth, character of material, and to get some idea of its gold content at depth. Results from the four preliminary holes enhanced the overall prospect and indicated a need for further sampling.

At this point a rectangular grid in the form of 800' x 1600' rectangles was laid out and holes were drilled at each intersection. This initial wide-spaced drilling over a 1500-acre tract was designed to determine the approximate size, shape, and possible value of the deposit. Using results from this drilling as a guide, the prospect data were further refined by drilling intermediate holes along some of the 1600' sides and, in turn, a selected area was divided into 400' blocks by intermediate drilling along the 800' lines. Each step, progressively taken, was designed to furnish the amount of information needed for evaluation at that stage. Had the results from any one step been unfavorable the project would have been dropped at that point. It should be apparent that a high degree of training and diagnostic skill is needed in this work.

In the foregoing example, a reliable placer valuation was based on a sampling density of one hole for each 3 1/2 acres of dredge ground which in this case was equivalent to a volumetric sample factor of about 1/500,000. For comparison, there are cases in which one hole for each 10 acres was found adequate and some in which no amount of drilling would suffice.

In summary: We can seldom, if ever, predetermine an optimum sampling pattern or the total number of samples required to evaluate a placer. For this reason initial sampling programs should be tentative or at least should be flexible enough to accommodate any changes dictated by the work as it progresses. In any case, detailed placer sampling is expensive and for this if no other reason, even the most comprehensive sampling program should progress from a simple beginning to its final form in a series of carefully evaluated steps.

Also, it is important that a distinction be made between sampling for the purpose of initial valuation and sampling to block out a finite parcel of mining ground. It sometimes takes only a few judiciously selected samples to determine that a property has no economic value while, on the other hand, an area showing good potential may require a thorough sampling in order to determine the value and other information needed to plan a successful mining operation.

Parenthetically, it should be pointed out that there is no final proof of the accuracy of placer sampling because in an ensuing operation, the mining and metallurgical losses can never be fully identified or measured. This is particularly true in the case of gold dredging.

TABLE I. The effect of a single small gold particle on placer samples of typical size.
The values shown are those which would result from one gold particle in a 1-foot sample increment.
Values shown are based on gold weights determined by the author, and gold at $35 per ounce.

Size of Drill Hole or Channel	SIZE OF GOLD PARTICLE And Effect on Sample		
	20-Mesh (6.57 mg.)	40-Mesh (0.91 mg.)	60-Mesh (0.27 mg.)
7½" dia.	58c/cu.yd.	8c/cu.yd.	2½c/cu.yd.
5¼" dia.	$1.18 "	16c "	5c "
3" dia.	$3.60 "	50c "	14c "
3" x 6"	$1.42 "	20c "	6c "
6" x 6"	71c "	10c "	3c "
6" x 12"	35c "	5c "	1½c "
12" x 12"	18c "	2½c "	3/4c "
16-inch pan 1/	$1.18 "	16c "	5c "

1/ @ 180 pans per cubic yard.

3. SAMPLING METHODS

a. *Existing exposures*: During the early stage of a placer investigation a quick answer may be needed to determine if a significant expenditure of time or money is justified for further investigation. At this stage, practical considerations may limit initial testing to a few samples taken from existing exposures such as creek or dry wash banks, road cuts, old mining pits, etc. These can be informative if properly used.

Existing exposures are usually tested by panning, particularly where the exposure is small. In most cases the bedrock will not be exposed and the distribution of available sample points will be far from ideal and, in either event, it should be obvious that extreme care is needed when evaluating the sample results. On the other hand, the only preparation needed may be the cleaning away of sloughed or weathered material to expose fresh surfaces and, because of this, the use of existing exposures offers a cheap, fast method for preliminary testing.

By themselves, small samples obtained from existing exposures can seldom be expected to indicate the actual value of the ground. They may, however, prove or disprove the presence of gold and, if correctly interpreted, they can indicate the range of values to be expected. Nevertheless, many reports intended to prove or disprove the *actual value* of placer lands are based on a few pan-size samples taken from existing exposures and offered at face value. Such reports and sample data must always be viewed critically and accepted with reservation until proved valid. The very offering of this type of data, without qualification, may be good reason to question the expertness or intent of the vendor.

In brief, samples from existing exposures taken in conjunction with a placer investigation can, if properly evaluated, provide valid, useful data. But, for the inexperienced or the unwary sampler, existing exposures offer an inviting and sometimes unrecognized trap.

b. *Hand-dug excavations*: Hand-dug excavations in the form of pits, trenches or small shafts are generally suited to dry, shallow ground. They are most effectively used to depths of about 7 feet which is the depth from which material can readily be thrown with a shovel. Where practicable, they may provide the best means for sampling a placer. They generally do not require such close supervision as drilling nor do they require the high cost, specialized equipment or highly trained personnel needed for placer drilling. An extra bonus is offered by hand-dug excavations in cases where they remain open for some time, permitting resampling at a later date. In the case of dredge prospects, the principal point in favor of pits or shafts, where applicable, is the fact that they provide a much better idea of boulder content and gravel sizes than do small-diameter drill holes. Boulders are always an important factor in dredging operations.

The simple equipment needed for hand-dug sample pits or shafts is a particular advantage in remote areas or where only a limited amount of sampling is anticipated. A few strategically located pits or shafts may show that the ground is entirely unfit for mining or that it will not warrant the expense of drilling. On the other hand, if they return good prospects and the ground is subsequently drilled, the expense of preliminary hand work will be but a small part of the overall cost. Successful mining companies learned long ago to approach placer prospects cautiously and to make a few diagnostic tests before putting their money in an expensive sampling program. In many cases hand-dug excavations adapt well to this need.

The use of hand-dug shafts for general placer sampling was common practice at the turn of the century when labor was cheap and experienced shaft men were readily available but, today, because of the high cost of labor and a scarcity of experienced men, they are seldom used on a wide scale. But they may still be used to advantage in special situations; for example, by prospectors who expect to find good bedrock values in ground too deep for effective use of pits or trenches. In such cases one or more shafts will usually be put down to test the gravels lying on bedrock. This approach is common where the owner or a promoter's main concern is to convince someone that the property should be more thoroughly prospected by drilling or other means.

Hand-dug shafts are commonly used to check selected churn drill holes and in many cases they are essential to correct interpretation of the overall drilling results. This is taken up in the section dealing with sampling by means of churn drills.

Most engineers agree that where feasible all of the material taken from a test pit or shaft should be washed. If the ground stands well and the walls of the excavation are cut square and parallel, it is an easy matter to determine the in-place volume of material removed and to directly compute its value after washing. An excellent description of a sampling program employing this procedure is found in an article by Sawyer (1932, pp. 381-383). The use of hand-dug shafts to prospect ground worked by the Wyandotte Gold Dredging Company of California has been described by Magee (1937, p. 186). Where an excavation cannot be measured accurately, it will be necessary to weigh the sample material or to measure its loose volume in a box or other container. In either case the indicated gold value will have to be converted to in-place or "bank" value by use of suitable conversion factors.

Where it is not practicable to wash all material, a good approximation of the ground's worth may be made by washing several pans per foot of depth as the excavation advances. But, this method is always risky unless the man in charge is completely impartial in his selection of material for panning and unless he is a trained placer sampler capable of applying experience-based judgement to the findings.

A somewhat similar procedure but one in which the personal factor is in part eliminated may be carried out as follows: Carefully deepen the pit or shaft in uniform, measured drops of say one foot at a time. Just prior to each deepening, carefully remove a sample by digging *below* the bottom of the pit or shaft to a depth exactly equalling the distance the bottom is to be dropped during the next deepening step. The minimum-size sample that can be conveniently taken in this manner will be one having a 12" x 12" cross section but if many rocks are present it may be somewhat larger. A sheet-iron caisson 18 or 24 inches in diameter by about 12 inches long can be used to advantage when the bottom sample must be taken from wet or ravelling ground. This type of sample may be taken from the center of the pit or shaft bottom or from any suitable place but once started, its location should remain fixed. By following this procedure, a progressive sample can be obtained from undisturbed material *ahead* of the main excavation.

Where the excavation walls stand reasonably well, a sample can be obtained by cutting a vertical channel up one side of the pit or shaft. The excavated sample material can be allowed to fall on a canvas placed at the bottom of the pit or it may be caught in a bucket or box held close to the point of cutting. Where conditions permit taking a true channel, the sample volume can be determined by direct measurement but in any case it is best to weigh each sample as a check.

The first step in channel sampling is to clear away all foreign material from the ground surface above the place to be sampled. Next, thoroughly clean the sample area by scaling down the face to be cut. Following this, remove any loose material from the bottom of the pit where it might interfere with the sampling operation and, finally, check the bottom for bedrock. A piece of heavy canvas or a tarp is then spread at the bottom of the face to be sampled and this is positioned to catch all material falling from the sample cut. Starting at the bottom, a uniform rectangular channel is cut upward to the top of the prospect pit or in some cases, to the top of a particular formation or sample increment. When starting a sample, the cuttings are usually allowed to fall on the canvas but when sufficient room becomes available, it is best to catch them in a suitable box or metal container held close to the point of cutting. As the sample channel advances upward, the cuttings are periodically transferred to one or more sample sacks which should be kept in the sampler's possession or under surveillance throughout the sampling operation. In most situations an ordinary prospect pick will be found satisfactory for cutting channel-type placer samples.

How large should the channel cut be? There is no simple answer because like other aspects of placer sampling there may be special factors to consider in each case. Speaking generally, however, it will be found that experienced placer samplers usually select a size somewhere between 3" x 6" and 12" x 12" in cross section but, contrary to popular belief, there is no minimum or optimum size which by itself will insure an adequate sample.

In ground where the water level is several feet below the surface, it is sometimes possible to sink pits or small shafts to bedrock using only buckets or a small pump to remove the water. These wet excavations are usually difficult to keep squared up but, in any case, some form of ground support may be needed to eliminate the working hazard and to minimize sampling errors which result when inflowing water carries values into the pit. A simple, easily-framed style of cribbing suitable for most shallow work is shown in Figure 3.

In loose, water-logged ground it may be impracticable to sink prospect shafts without the use of telescoping steel caissons. Certain problems and special procedures associated with caisson work have been discussed by Steel (1915, pp. 66-68) and Dohney (1942, pp. 48-49).

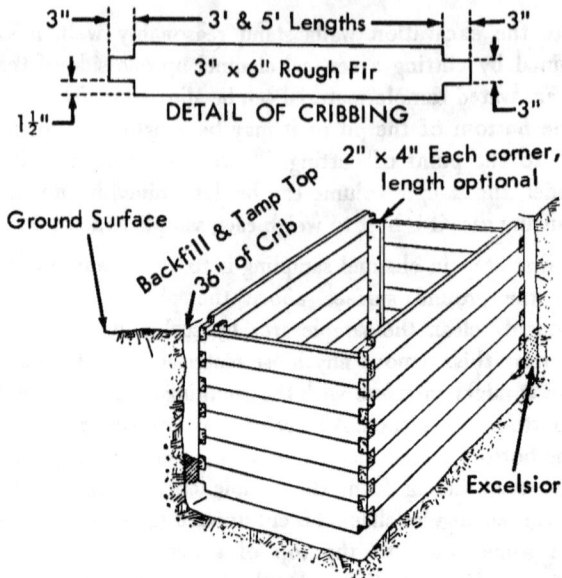

```
3"→|  |←———— 3' & 5' Lengths ————→|  |←—3"
 ↑ ┌─┐                                   ↓  ┌─┐
 ↕ │    3" x 6" Rough Fir              │  │  │↑
1½"─┘   DETAIL OF CRIBBING            └──┘  └─3"
```

FIGURE 3. — Test pit cribbing suitable for use in shallow placer ground. (From U.S. Bureau of Reclamation Concrete Manual, 7th edition, 1963).

At first glance the use of hand-dug pits or shafts would seem to be an effective inexpensive way of prospecting shallow placers but this approach is too expensive for general use today, where a high dollar value must be placed on time. Nevertheless, hand-dug excavations intelligently used can serve a useful purpose in that, under favorable conditions, a few well located pits or shafts may indicate the value of a prospect before any great expenditure of money is made.

c. *Machine-dug shafts:* Machine-dug shafts have been found both useful and economic in a number of instances but they are not a cure-all for placer sampling problems. The equipment used can broadly be divided into two types: rotary bucket drills and clamshell-type excavators. To be effective, either type must be capable of doing three things: (1) dig a hole large enough for a man to enter and inspect the ground or to cut samples from the shaft walls; (2) have a reasonably fast digging rate; and (3) be self-contained, freely movile and capable of negotiating rough terrain.

Truck-mounted bucket drills have been in use for many years by excavation contractors, particularly those specializing in foundation test work. Their equipment is usually designed to dig a hole 24 or 36 inches in diameter and to depths of 60 feet or more. The digging unit can best be described as a straight-sided bucket which has two radial openings in its bottom and excavating blades extending below the openings. Rubber flaps fitted over the openings allow cuttings and rocks to 6-or 8-inch size to pass into the bucket but seal the openings when the loaded bucket is being hoisted. The bucket is rotated by means of a square "Kelly" bar which, in turn, is rotated by a ring gear on the drill rig. See Figure 4.

FIGURE 4. — *Rotary bucket drill making a 24-inch diameter hole in desert placer material. The average drilling rate for 90 foot holes was about 70 feet per shift. Progress was slowed by the presence of numerous hard, caliche-cemented layers.*

Attempts to adapt this type of drill to placer sampling have met with mixed success but speaking generally, rotary bucket drills have been found useful where the ground is dry and firm and where there are few rocks too large to pass through the bucket openings. On the other hand, they have been found poorly suited to the hard, rocky ground encountered in many gold placers.

The successful use of rotary bucket drills to sample dry placers has been described in articles by Draper (1932, p. 537) and by Prommel (1937). Prommel describes a sampling project in which 1,239 test holes 28 inches in diameter were drilled in 147 days. Their combined depths amounted to 16,705 feet. The cost of this work using two rotary bucket drills was 98.6 cents per foot of hole but this did not include engineering or sample testing charges and, it should be noted that the low cost reflects depression-level prices of 1935.

Two clamshell-type machines designed specifically for placer prospecting are: An air-actuated unit sold under the trade name KLAM, and a cable-actuated unit known as the PAR-X Placer Sampler. The KLAM is essentially a manganese steel clamshell bucket of 1 cubic foot capacity, fitted with replaceable teeth and actuated by compressed air. The digging unit consists of the bucket, an air cylinder placed directly above the bucket and a 15-foot stem as shown in Figure 5. The combined digging unit which weighs one ton is suspended from the mast of a drill truck by means of a wire line spooled on a suitable hoist. According to the manufacturer, the standard unit digs a round, vertical hole 24 inches in diameter. and is efficient to a depth of 100 feet. It is said to dig from 2 to 20 feet per hour depending upon depth, formation, and on whether or not casing is used.

The PAR-X machine differs from the KLAM in that the clamshell is attached to a telescoping stem or bucket carrier which limits the depth of hole to about 30 feet. The opening and closing action is controlled by wire ropes. The unit is truck-mounted and, according to Huttl (1941, pp. 55, 56), its digging speed averages 35 feet per day in normal ground. It digs a hole 28 inches square without casing.

Machine-dug shafts are best suited to bulk sampling, that is, sampling in which the entire volume of material removed from the shaft is run as a sample. Where the ground stands well channel samples can be cut from the shaft walls but there is hardly room for a man to work effectively in a 24-inch diameter hole. However, the fact that machine-dug shafts can be entered for inspection is a point in their favor, particularly in new fields where little is known about the sub-surface conditions.

Something proponents of this method of placer sampling seldom mention is the fact that an uncased, machine-dug shaft passing through alternating hard and soft layers or through ravelling ground will have a non-uniform diameter, that is, it may be appreciably larger in the softer horizons than in the hard. This means that where over-diameter shaft sections occur in ground that is above average or below average mineral

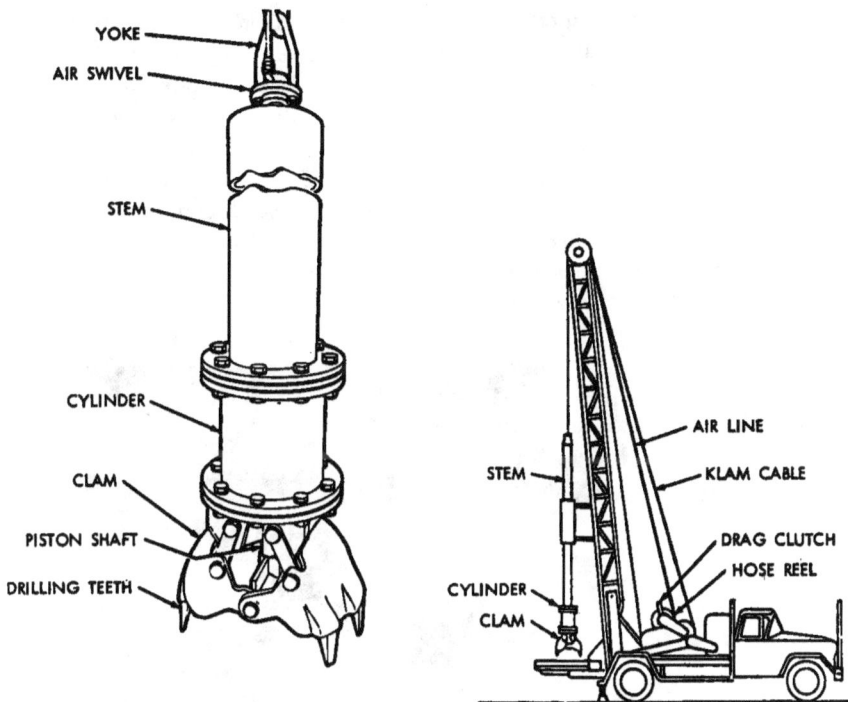

FIGURE 5. — *Air-actuated excavator manufactured under trade name of KLAM. The unit shown is designed to dig holes 24 inches in diameter to depth of 100 feet or more.*

value, the end result will be an unintentional salting or an impoverishment of the bulk sample.

To cite an actual case, a dry, desert placer was sampled by means of 24-inch diameter machine-dug shafts ranging from 60 to 90 feet deep. Both rotary-bucket and Klam-type excavators were used. The valuable mineral was magnetite occurring in the form of black sand. When the completed shafts were entered for inspection, it was found that the shafts had passed through a series of hard, caliche-cemented gravels separated by relatively loose sand and gravel strata and, in all cases, the shaft diameters were appreciably larger in the softer materials than in the cemented horizons. Sections were found enlarged to as much as 2½ times the normal 24-inch diameter. If the grade of material throughout the shaft had been uniform this local enlargement would be of no consequence, but it was found that the enlarged sections were often in magnetite-rich sands. As an end result, samples (representing 30-foot shaft increments) contained too large a proportion of the richer material which in effect amounted to unintentional salting. Had this been a gold prospect the errors caused by non-uniform hole size would have been substantial.

d. *Backhoe excavators*: The development of compact, tractor-mounted backhoe excavators in recent years has given the mineral examiner a tool which adapts well to many of his placer sampling needs. These go-anywhere, high-performance machines dig, load and backfill. The combination most frequently employed is a hydraulically-operated backhoe unit mounted on the back end of a rubber-tired tractor plus a scraper or loading bucket at the front end. See Figure 6. A typical rig will dig 10 feet deep, reach more than 15 feet from the pivot point and will load into a truck 11 feet high. The backhoe bucket has more than

FIGURE 6. — *Backhoe excavator digging trench in placer ground. Hand-dug channel samples were later taken from the trench walls.*

7,000 pounds digging force and generates a pryout force of 15,000 pounds or more. Most placer ground is handled with ease. A 1 1/2-yard, track-mounted backhoe (Thew-Lorain) has a depth capacity of 27 feet and is available with a range of buckets that provide cutting widths from 12 to 38 inches.

In most areas, this type of equipment is available on a rental basis from contractors who will be found listed in the yellow pages of the telephone book under "excavating contractors." Rental rates vary with locality and machine size but $10 to $14 per hour is not unusual. This price includes the operator. On short-term jobs the charge starts at the time the equipment leaves the contractor's yard so it is well to have the work well planned and laid out in advance.

In connection with a placer sampling job near Grants Pass, Oregon, in 1966, thirteen shafts ranging from 4 to 12 feet deep were dug using a tractor-mounted backhoe equipped with a 38-inch "graveyard" bucket. The backhoe with operator rented for $11 per hour. The 13 shafts were dug in 6 hours and the travel time was two hours bringing the total charge to $88. This amounted to a little less than $1 per foot of shaft and provides a good example of the cost and work accomplishment that can be expected when using this type of machine.

In most placer gravels the walls of backhoe-dug pits or trenches stand well and can be sampled by cutting a vertical channel in the face of the standing material. Depending on requirements and conditions, the sample channel can be cut with the backhoe bucket (usually 24" wide) or by hand. Where the presence of much large rock or ravelling ground precludes the use of channel cuts, either all excavated material or alternate buckets can be set aside as a bulk sample. See Table III, p. 99.

The tractor-mounted backhoe is probably the most versatile placer sampling tool available to the mineral examiner today.

e. *Bulldozer trenches*: The use of bulldozers to dig trenches for prospecting shallow placers is too common and too well known to need more than passing mention.

Because of its design, the bulldozer is more effective as a pusher than as a digger and for this reason it is best suited to ground less than 10 feet deep and to a working (pushing) distance not exceeding 200 feet. Two types of blade control are in general use; the blade may be raised and lowered hydraulically or it may be operated by a cable control. For digging prospect trenches the hydraulic blade is considered best as it provides a positive downward pressure; in fact, it allows much of the tractor's weight to be put on the blade while digging hard ground. The cable-controlled blade has only its own weight to hold it down and as a result its cutting edge has a tendency to ride over any hard spots in the bottom of the trench.

The cost of bulldozer-dug trenches depends so much on the type and size of machine used and on the depth and hardness of the formation, and the operator's skill; any estimate of working speed or costs must be made on a case-by-case basis. Lorain (1944, pp. 1-8) has tabulated field data for trenching with bulldozers but his 1944 cost figures cannot be used without updating.

Possibly the greatest advantage to be gained in prospecting placers by means of bulldozer trenches, where applicable, is that they permit first-hand inspection of the ground. This makes it possible to visually determine the character and size of gravel to be dealt with and, in addition, the pay or barren sections can sometimes be found by simple inspection and panning of the trench walls. Another feature is that a considerable area of bedrock can be exposed for inspection or testing as compared to pits or drill holes. In many cases, even where the ground is more than 10 feet deep, the results obtained from a few days of trenching with a bulldozer will provide the information needed to decide whether or not a more extensive (and expensive) sampling program is warranted.

f.   *Sampling with churn drills*: Drilling is usually resorted to in deep or wet ground where sampling by means of pits, trenches or shafts is not practical. Many western gold placers, particularly those exploited by bucket-line dredges, have been sampled by churn drilling and it can be said generally that this method adapts well to the dredge operator's needs.

The basic drilling equipment consists of a heavy casing, a chisel-shaped bit suspended from a drill line and a vacuum-type sand pump. The overall operation can be divided into five steps as follows:

Driving.

Drilling.

Pumping.

Panning.

Pulling and moving.

The first step provides a measured core of gravel which serves as the sample. The gravel core is obtained by driving the casing (commonly referred to as the drive pipe) into the ground by striking it with a driving block attached to a heavy drill stem. The bottom end of the drive pipe is fitted with an alloy-steel shoe made with a sharp, beveled cutting edge. The effective diameter of this beveled edge determines the area of gravel or sample cut by the drive shoe as it is forced downward. A second step breaks up the core and prepares it for removal from the pipe. This is done by chopping the core with the drill bit which is repeatedly raised and dropped by means of a walking beam or crank acting on the drill line. Water is added during the drilling operation and when the material is reduced to a size that will permit pumping, the resultant sludge is removed from the hole with a special vacuum-type sand pump. The cuttings are transferred to a measuring bucket which serves as a check to indicate if excess core is being removed or if extraneous material is running into the pipe.

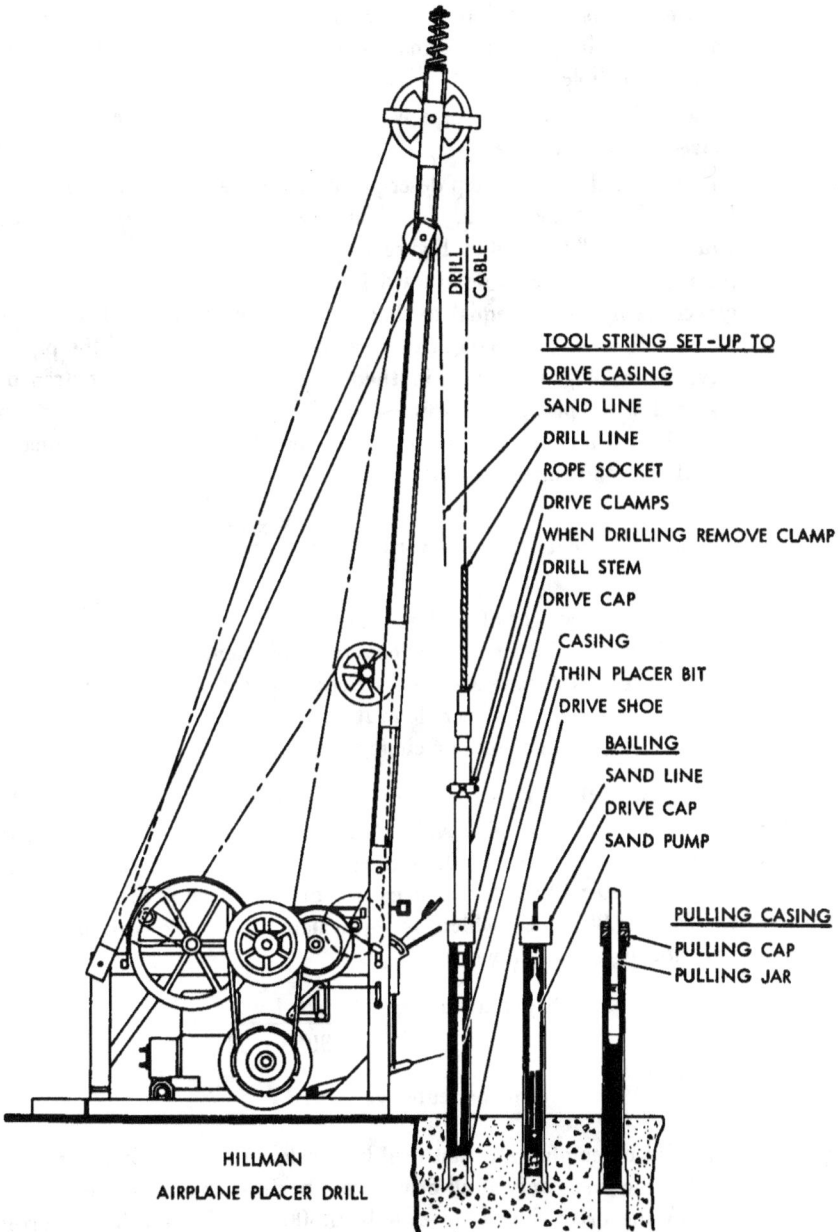

DRILL CABLE

TOOL STRING SET-UP TO DRIVE CASING
SAND LINE
DRILL LINE
ROPE SOCKET
DRIVE CLAMPS
WHEN DRILLING REMOVE CLAMP
DRILL STEM
DRIVE CAP
CASING
THIN PLACER BIT
DRIVE SHOE
BAILING
SAND LINE
DRIVE CAP
SAND PUMP
PULLING CASING
PULLING CAP
PULLING JAR

HILLMAN
AIRPLANE PLACER DRILL

FIGURE 7. — Basic equipment employed in placer drilling. (Courtesy of C. Kirk Hillman Company, Seattle)

45

The cuttings are washed in a rocker and the rocker concentrate further reduced by panning. The panner counts the gold colors recovered from each drive, estimates their weight and enters this information in the drill log. Examples of placer drill logs will be found in Appendix D.

The driving, drilling and pumping steps are successively repeated for each drive (usually 1 foot at a time) until the hole is completed. The engineer in charge then weighs the combined gold and calculates the value of the hole.

As a final step, the drive pipe is recovered by pulling. The drill is then moved to the next hole.

An illustrated, step-by-step description of the placer drilling process will be found in ,Appendix D,, to which the reader is referred. The basic equipment is illustrated by Figure 7.

As the pipe is driven downward it will, in theory, cut a cylinder of gravel having an area equal to that of the drive shoe and a length equal to the drive. A short length of core left in the bottom of the pipe to serve as a plug will, also in theory, prevent run-ins of extraneous material when pumping. Assuming ideal conditions and a perfect core, the value of the hole (or the sample increment) would be determined by use of the following equation:

$$\text{cents per cubic yard} = \frac{W \times V \times 27}{A \times D}$$

in which W = weight of gold in milligrams.
V = value of gold in cents per milligram.
A = effective area of drive shoe in sq. ft.
D = length of sample in ft.
27 = cubic feet in a cubic yard.

To illustrate the above equation, assume a 32-foot hole drilled with a standard 6-inch drive pipe and 7 1/2-inch (outside diameter) drive shoe with an effective area of 0.3068 sq. ft. Also assume a recovered gold weight of 50 milligrams. As a matter of convenience the gold will be considered 887-fine, which is equivalent to 1/10c per milligram. Substituting we would have:

$$\text{cents per cubic yard} = \frac{50 \times 0.1 \times 27}{0.3068 \times 32}$$

$$\text{cents per cubic yard} = 13.7$$

It should be emphasized that drill hole values calculated in this manner would be correct *only* if all drives were perfect, that is, if the effective area of the drive shoe were exactly 0.3068 sq. ft., and if the correct length of core had entered the pipe with each drive. In the preceding example, the 50 milligrams of gold would have been obtained from a cylinder of gravel exactly 7 1/2 inches in diameter by 32 feet long.

But in practice a perfect core is rare and we find the measured core rise and the volume of material recovered from each drive to be greater or less than their theoretical amounts. When drilling gold placers, any deviation from the norm is important because when basing value calculations on small-diameter holes, any deviation between the theoretical sample size and the actual sample size becomes critical. For this reason, any excess or deficiency of core should be taken into consideration during the calculation procedure and suitable corrections applied.

Erratic drill cores are caused by a variety of things. A rock partially blocking the drive shoe may move downward with the pipe and force material beneath it to one side. The result would be a deficient core. A normal-length drive in tight ground often gives less core than it should. This is caused by material packing in the pipe and as it becomes more difficult for the core to enter, some will be pushed aside. The experienced driller sometimes prevents this by drilling a short distance ahead of the shoe. He should not, however, "drill ahead" without permission of the engineer in charge because there are cases where drilling below the shoe creates more problems than it solves.

When a drive is made through tailings or through loose, rocky ground, the initial core rise may be normal but subsequent pounding by the drill bit tends to push the core material back down the pipe and out of the hole. As a result, the pump may get little or no core recovery. In loose, wet ground there is always danger of overpumping, that is, pulling excess material in from under the drive shoe. In "running" ground the driller should keep the drive pipe full of water to offset outside ground water pressure (which tends to push loose material into the pipe) but in spite of such precaution, excess cores can be expected.

Ideally, a 1-foot drive should produce a 1.7-foot core rise; this 0.7-foot increase reflecting the difference in diameters between a 6-inch drive pipe and its 7 1/2-inch shoe. But in practice we rarely get the theoretical amount of core. Instead, we get too much or too little for a given drive. Now, the most important thing to remember is that if a placer drill sample is to be meaningful it must take into account the amount of material *actually* obtained from the sample point in question. Stated simply, if a drillhole increment contains too much core its value will have to be downgraded. If it contains too little core its indicated value may be upgraded. Most placer engineers base their adjusting procedures on either the ratio between theoretical and *actual* core rise, or theoretical and *measured* core volume. Depending on which is used, the basic formula would be:

corrected gold weight = $\dfrac{\text{theoretical core rise}}{\text{actual core rise}}$ x estimated [1] weight

---

[1] In practice the corrections are usually applied to the estimated gold weight for each drive. The individual corrections are carried forward for use when calculating and end value for the completed hole.

$$\text{corrected gold weight} = \frac{\text{theoretical core volume} \times \text{estimated weight}}{\text{actual core volume}}$$

In general practice both core rise and volume corrections will be considered, but the most conservative correction will be used.

Experience has shown that adding large plus corrections at their full value is likely to result in an overestimate and for this reason some engineers limit any "corrected" figure to an amount not more than double the initial uncorrected figure. They will, however, apply the full amounts of all minus corrections, this being considered a form of safety factor. Actually, in the case of gold placers, there are no universal rules for interpreting or "correcting" the samples obtained from churn drilling. Validity of the end result will depend largely on insight and the experience of the engineer in charge.

Many of the details to be considered when logging placer drill holes and processing the data have been given by Harding (1952, pp. 96-98) and Daily (1962, pp. 80-88), to whom the reader is referred. From these it will be seen that placer drilling requires specially trained personnel and that "correcting" the drill results is a specialty that is not within the experience range of most mining engineers. For this reason, any placer valuation based on drilling should be viewed critically until the qualifications of the engineer in charge have been established. This is particularly true where the original drill logs are not available for inspection, or where drilling cannot be compared against past production from the particular deposit or area.

Something should be said about the so-called "Radford" factor [1] which, when used, applies an arbitrary plus correction to the drill hole calculation. Early placer drillers soon became aware that their actual core recoveries were something different than the theoretical amounts and it was reasoned that because of progressive rounding of the cutting edge, a 7 1/2-inch drive shoe will in normal use cut something less than 0.3068 sq. ft. of material. Mr. Radford, on the basis of his experience and observations, concluded that the sample taken in by a 7 1/2-inch shoe driven 1 foot will, on an average, be 0.27 cu. ft., or as more commonly stated, 1/100 of a cubic yard per foot of drive. This figure was widely accepted by the early drillers.

The Radford factor is applied by substituting 0.27 for "A" in the basic equation where cents per cu. yd. =

$$\frac{W \times V \times 27}{A \times D}$$

---

[1] Also known as the Keystone Constant.

Using our earlier example this equation would become:

$$\text{cents per cubic yard} = \frac{50 \times 0.1 \times 27}{0.27 \times 32}$$

cents per cubic yard = 15.6

The first calculation using the theoretical drive shoe area of 0.3068 sq. ft. for "A" gave a value of 13.7 cents. Substituting 0.27 gave a value of 15.6 cents. It is seen that use of the Radford factor increased the calculated value by 12 percent.

In some cases other factors were applied to decrease rather than to increase the calculated value. Usually, the Radford or other factor was applied to the entire hole, that is, they were not applied on a drive-by-drive basis. It should be obvious that the use of any arbitrary correction factor is hazardous because ground character will vary in different properties and for that matter in different parts of a single drill hole.

Today, considerable difference of opinion exists concerning the validity or use of the Radford factor. It has been mentioned here to alert readers who may not be familiar with placer drilling that many of the placer valuations found in reports, particularly in old reports, have been based on calculations employing the Radford or other arbitrary upgrading factors.

In arctic regions, permanently frozen ground is drilled without casing with exception of a short length at the surface to serve as a collar and tool guide. Sample volumes are determined by measuring the displacement of known quantities of water poured into the hole. Arctic drilling and the special procedures employed have been well described by Doheny (1941, pp. 47-49).

Most placer-type churn drills used in the United States are powered by gasoline engines and designed for use with 6-inch drive pipe and tools. Typical arrangements are illustrated by Figures 8 and 9. Drills in common use are the Keystone Model 70 [1], a truck-mounted unit for use with 6-inch pipe; the Keystone Model 71, a 6-inch traction drill mounted on a crawler base; the BE 22-T [2], a crawler-mounted 6-inch drill; the Hillman Prospector [3], a 6-inch truck or crawler-mounted rig and the Hillman Airplane Drill, a compact, skid-mounted machine designed for use with 4 or 5-inch drive pipe. A crawler-type traction drill complete with 40 lengths (about 240 ft.) of drive pipe and the necessary tools will weigh 10 to 15 tons.

---

[1] Now called "Speedstar"; manufactured by Buffalo-Springfield Co., Enid, Oklahoma (successor to Keystone Driller Co.).

[2] Manufactured by Bucyrus-Erie Co., South Milwaukee, Wisconsin.

[3] Manufactured by C. Kirk Hillman Company, Seattle, Washington.

FIGURE 8. – Truck-mounted Keystone Model 70 placer drill.

Hand-powered drills are widely used in foreign fields and regions where access is difficult and labor cheap. The Banka or Empire [1] drill employs a rotating casing equipped with a serrated cutting shoe. The Ward [1] hand-powered drill employs 4 or 5-inch drive pipe and standard tools.

A drill crew usually consists of the driller who is responsible for operation of the drill, a helper to assist with general work and handling the tools and a panner who must process the samples and keep an accurate log of the hole as it progresses. A typical placer drilling set-up is shown in Figure 10. A fourth man may be required where water must

[1] Available from New York Engineering Co., New York, New York.

FIGURE 9. — *Hillman 4-inch Airplane drill. This compact, skid-mounted machine is widely used for reconnaissance-type prospecting.*

be hauled or drill sites prepared. An experienced driller has become so accustomed to the sound of drilling and the feel of the drill line, he can usually tell just what is happening in the hole. By keeping careful watch and logging all pertinent data, he provides the information needed by the engineer for "correcting" the indicated values. The panner who should have some technical as well as practical training, often serves as the crew foreman.

As a rule, when prospecting narrow, stream-type deposits the holes are drilled at relatively close intervals along lines laid out at right angles to the general trend of the deposit as shown in Figure 11. Hole spacing

FIGURE 10. — *A typical placer drilling set-up. Note rocker for washing samples.*

along the line is commonly 100 feet or less, while the distance between drill lines is greater, often on the order of 500 feet or more. This type of drilling pattern is designed to search out and delineate relatively narrow pay channels.

In widespread placers, particularly those having a somewhat uniform mineral distribution, holes are usually put down at the intersections of a rectangular grid pattern designed to cover the deposit as shown in Figure 12. The drilling ratio will vary from one hole for every acre to one hole for every 2, 3 or possibly 10 acres, depending on the distribution of mineral, previous experience in similar ground and other factors. In the Folsom, California dredge field the average hole density was about 1 hole for every 4 acres. On the other hand, there are placers which cannot be adequately sampled by any amount of drilling.

Although a regular drilling pattern may be preferred, there are situations in which an irregular drill pattern should be employed. In any case, it is important to avoid placing the holes in a way that would tend to exaggerate either the higher or the lower-grade areas. The selection of a drilling pattern and the initial placement of holes is normally a responsibility of the engineer in charge.

Assuming a completed drilling project with favorable results, the next step is to make an overall evaluation. To calculate the yardage and the average value of a deposit which has been drilled on a grid pattern, the drill holes are first connected by imaginary lines to divide the deposit

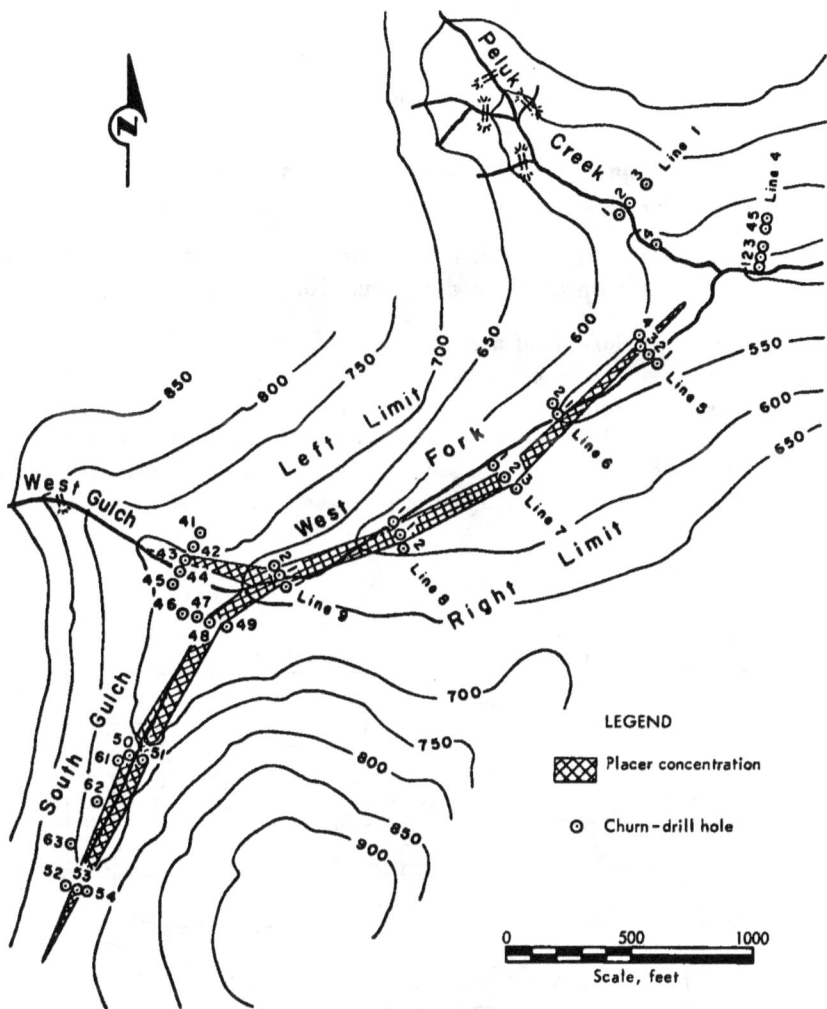

FIGURE 11. – *Typical drilling pattern for prospecting narrow, stream-type deposits.* *(From USBM R.I. 6587).*

into blocks or triangles. See Fig. 13, page 56. The volume and value of each triangle or block is then calculated using weighted drill holes values and standard ore reserve calculation procedures. The procedural details will not be set out here as they are well known to most field engineers and have been adequately covered in the technical literature. Daily (1962, pp. 86, 87) and Doheny (1942, Part II, pp. 43, 44) have described the applications to placers. In principle, the value of each block or triangle is calculated by the following formula:

53

$$V_a = \frac{\text{Sum of products (d x V)}}{\text{Sum of depths (d)}}$$

Where V_a = average value of block or triangle expressed in cents/cubic yard.

d = depth of each bounding hole in feet.

V = prospect value of each hole in cents/cubic yard.

A value for the deposit is subsequently arrived at by combining the individual triangles as shown in Figure 13.

For the valuation of narrow, stream-type deposits where the holes are drilled at close intervals, say 50 or 100 feet apart on relatively far-spaced drill lines, a calculation based on the mean-area formula is usually employed. This relies on a formula used for making cut and fill estimates in road work in which it is inferred that the volume of a prism having somewhat similar end sections can be expressed by the equation:

$$\text{Volume} = \frac{A_1 + A_2}{2} \times L$$

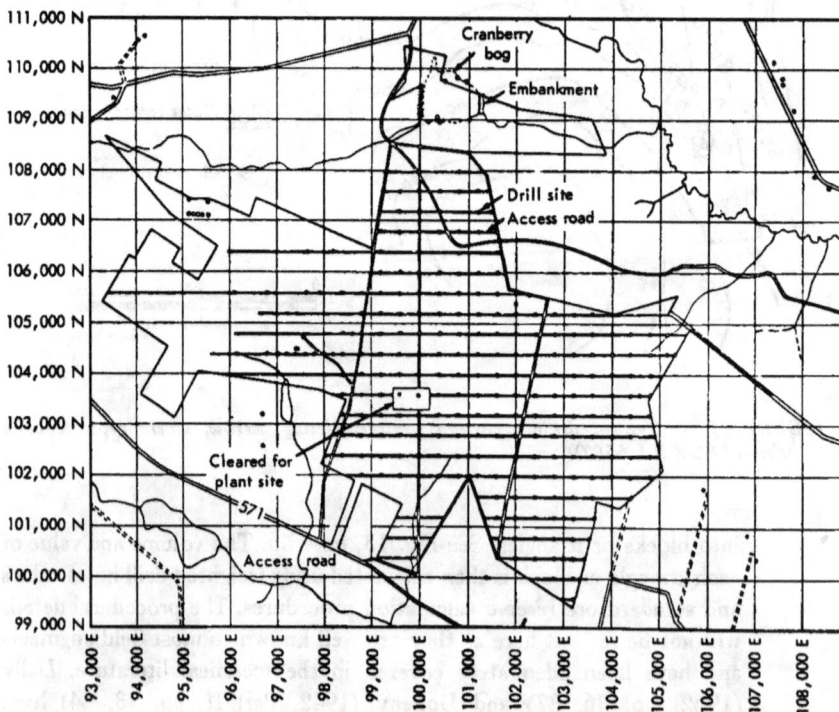

FIGURE 12. – An example of holes drilled on a rectangular grid to sample a heavy-mineral deposit. In this case, holes are on the corners of 400-foot squares that compose the drilling grid. (From USBM I.C. 8197)

Where

A_1 = the cross sectional area of one end of the block.

A_2 = the cross sectional area of the opposite end.

L = distance between ends.

Applied to placer valuation, the average prospect value of the two end sections of a gravel block is calculated and this value applied to the block volume derived from the foregoing equation. See Fig. 14, page 58.

When computing the average prospect value of each end section it will, of course, be necessary to consider the area of influence exerted by each drill hole. At the risk of oversimplification, a commonly used procedure can be described as follows:

Consider an elongated block of gravel bounded on its two ends by drill lines containing relatively close-spaced holes. The cross-sectional area between each pair of holes is found by multiplying their average depth by the distance between them. Multiplying this area by the weighted prospect value of the two bounding holes gives an area x value figure. This procedure is repeated for each pair of holes across the line and the same is done at the opposite end of the gravel block. To find the average value of the two drill lines, divide the sum of area x value products (from both lines) by the sum of the individual areas. The resultant weighted prospect value for the combined ends is then applied to the entire volume between the drill lines to arrive at a dollar value for the gravel block. The procedure is illustrated by Figure 14.

When drilling a gold placer, two holes put down in close proximity will rarely check and, in fact, the divergence in value will probably be too great to be reconciled by the layman. This problem is not necessarily a reflection on the drilling itself but, instead, reflects the small sample area encompassed by a 7 1/2-inch drive shoe and the typically irregular distribution of gold in placers. Interpreting "check" holes is another of the many things in placer valuation which rely heavily on experienced-based judgement.

In a like manner shafts sunk over drill holes should be expected to show appreciable variations in value when compared with their respective drill holes. This is particularly true in loose formations or in old tailings where drilling conditions may be far from ideal. Check shafts put down over drill holes do, however, allow inspection of the gravel in place thus providing the examining engineer with a wealth of information not available from drilling alone which in turn results in a more accurate evaluation of the deposit.

The average rate and cost of the placer drilling are topics which have received little attention in the technical press, none of which is recent. This largely reflects a tendency for operating companies to keep their prospecting data confidential plus the difficulty in identifying hole-by-hole performance and cost data among general operating records.

Drilling rates when figured on a day-to-day or a hole-to-hole basis will be found to vary widely. The variations reflect many factors including moving and set-up time, actual drilling, stuck or lost tools, breakdowns, pipe pulling, etc. To be meaningful, any *average* figure must take all such contingencies into account, but few do. For those who must make preliminary estimates, a good experience-based figure to use is 10 feet per calendar-day over the life of a drilling project. This may appear

Step 1—GRAPHIC DISPLAY OF PROSPECT DATA

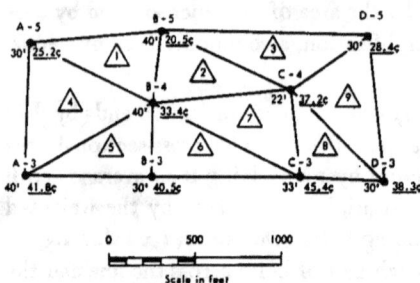

Scale in feet

Step 3—VALUE CALCULATION

TOTAL VOLUME = 2,012,305 cubic yards[2/]

GROSS VALUE = $684,065[2/]

$$\text{AVERAGE VALUE} = \frac{\$684,065}{2,012,305} \times 100$$

$$= 34.0c \text{ per cubic yard}$$

[2/] From Step 2.

Step 2—DETERMINE VOLUME AND VALUE FOR EACH TRIANGLE

Δ NO.	HOLE	DEPTH FT.	VALUE c	D x V	AVER. DEPTH	WEIGHTED VAL. c	AREA 1/ SQ. YDS.	VOLUME CU. YDS.	VAL. $
1	B-4	40	33.4	1336	36.6	26.52	20,500	250,100	66,326
	B-5	40	20.5	820					
	A-5	30	25.2	756					
		110		2912					
2	C-4	22	37.2	818	34.0	29.16	17,700	200,010	58,323
	B-4	40	33.4	1336					
	B-5	40	20.5	820					
		102		2974					
3	C-4	22	37.2	818	30.7	27.06	20,500	209,100	56,582
	B-5	40	20.5	820					
	D-5	30	28.4	852					
		92		2490					
4	B-4	40	33.4	1336	36.6	34.24	30,100	367,220	125,736
	A-3	40	41.8	1672					
	A-5	30	25.2	756					
		110		3764					
5	B-3	30	40.5	1215	36.6	34.43	16,800	204,960	78,766
	B-4	40	33.4	1336					
	A-3	40	41.8	1672					
		110		4223					
6	C-3	33	45.4	1498	34.3	39.32	19,650	224,010	88,080
	B-3	30	40.5	1215					
	B-4	40	33.4	1336					
		103		4049					
7	C-3	33	45.4	1498	31.7	38.42	22,850	242,210	93,057
	C-4	22	37.2	818					
	B-4	40	33.4	1336					
		95		3652					
8	C-3	33	45.4	1498	28.3	41.80	13,100	123,140	51,473
	C-4	22	37.2	818					
	D-3	30	38.3	1149					
		85		3565					
9	C-4	22	37.2	818	27.3	34.31	21,050	191,555	65,722
	D-3	30	38.3	1149					
	D-5	30	28.4	852					
		82		2819					
						TOTALS	182,250	2,012,305	$684,065

1/ Measured by planimeter.

FIGURE 13. — Use of the "triangle" method for calculating prospect values where deposits are drilled on a grid pattern.

56

conservative but it also reflects an actual *average* daily [1] footage obtained under a variety of drilling conditions in the United States, Canada and South America.

A word of caution. Where placer drilling is contracted and the contractor's payment is based on footage drilled, there may be a tendency to sacrifice accuracy for speed. Under such conditions the engineer in charge must be constantly alert and he should insist on having full authority over all aspects of the work.

The cost of placer drilling where no unusual problems are encountered will directly reflect the cost of labor. Because of the scarcity of qualified placer drillers and panners, drill crews are hard to find and, as a result, they can demand and usually receive premium wages. Other expenses include the purchase or rental of a drill rig, tools and supplemental equipment such as flatbed and pickup trucks, transportation to the job, supervision, engineering, insurance, fuel, supplies and incidental labor. Today, in the United States the minimum direct cost for operating a standard 6-inch placer churn drill on a one-shift basis will be about $125 per day, based on rental of the drill rig and tools at $50 per day and other expenses. Applying the average figure of 10 feet per calendar day over the life of a project, a minimum cost of $12.50 per foot of drilling is indicated but when incidental expenses are considered, the overall drilling cost may be nearer $15 per foot. A company operation utilizing permanent personnel or a contractor working for an extended period of time could quite possibly lower costs but the point being made here is that placer drilling today, if properly done, is expensive. If improperly done it is worse than useless, as shown by too many cases where inadequate drilling that was intended to save money led to the equipping of sub-economic deposits with resultant loss of large capital investments.

Speaking generally, the churn drill today is too expensive for casual or indiscriminate use as a placer sampling tool but it still offers the only feasible means for sampling some placers.

g. *Bulk samples*: Bulk samples may be called for when dealing with placers having an erratic distribution of values, especially where the pay is found as coarse gold or nuggets. Depending on conditions, samples containing a portion of a cubic yeard, or as much as tens or hundreds of cubic yards may be needed for a reliable estimate of value.

The so-called "plant run" is a form of large-volume bulk sample in which a portion of the deposit is mined and processed under actual working conditions. A pilot plant employing full-scale processing and recovery equipment is sometimes used. But when relying on plant runs, good judgement must be used to insure that the points or areas selected for large-scale sampling actually represent the deposit as a whole, because high costs usually limit this type of testing to a few points.

[1] One 8-hour shift per day.

One school of thought gives great importance to the supposed superior accuracy of bulk samples as compared to smaller samples such as those provided by churn drills or channel cuts. While bulk sampling per se is a valid method and bulk samples do, in many cases, offer the most satisfactory method of testing erratic, high-value ground, they are at the same time open to certain criticism. For example, bulk samples must be representative of the deposit, a feat not easily achieved where they are few in number. Another important consideration is the fact that all facets of a bulk sampling program cannot have personal supervision all of the time and, for this reason, salting can be perpetrated with comparative ease. It is important that all such possibilities be considered during the evaluation of bulk sampling results or when considering any actions based on this type of sampling.

Step 1—GRAPHIC DISPLAY OF PROSPECT DATA

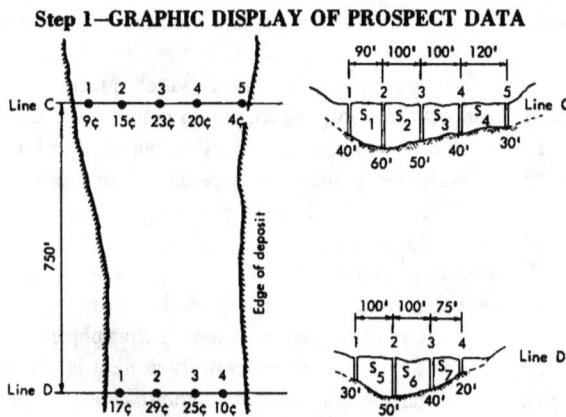

Step 2—DETERMINE SEGMENT AREAS

DRILL LINE	SEGMENT	BOUNDING HOLES	DEPTH FT.	AVER. DEPTH	WIDTH FT.	AREA SQ. FT.	AREA SQ. YDS.
C	S_1	C-1 C-2	40 60	50	90	4,500	500
C	S_2	C-2 C-3	60 50	55	100	5,500	611
C	S_3	C-3 C-4	50 40	45	100	4,500	500
C	S_4	C-4 C-5	40 30	35	120	4,200	466
D	S_5	D-1 D-2	30 50	40	100	4,000	444
D	S_6	D-2 D-3	50 40	45	100	4,500	500
D	S_7	D-3 D-4	40 20	30	75	2,250	250

FIGURE 14. — Use of the "mean area" method for calculating prospect values in narrow, stream-type deposits.

Step 3—DETERMINE WEIGHTED VALUES

DRILL LINE	SEGMENT	HOLE	DEPTH FT	VALUE	D x V	WEIGHTED VALUE	SQ. YDS.[1]	A x V
C	S_1	C-1	40	9c	360			
		C-2	60	15c	900	12.6c	500	6,300
			100		1260			
C	S_2	C-2	60	15c	900			
		C-3	50	23c	1150	18.6c	611	11,365
			110		2050			
C	S_3	C-3	50	23c	1150			
		C-4	40	20c	800	21.6c	500	10,800
			90		1950			
C	S_4	C-4	40	20c	800			
		C-5	30	4c	120	13.2c	466	6,150
			70		920			

AREA LINE C = 2,077

SUM A x V = 34,615

Average value of line C = $\dfrac{34,615}{2,077}$ = 16.7c per cubic yard.

DRILL LINE	SEGMENT	HOLE	DEPTH FT	VALUE	D x V	WEIGHTED VALUE	SQ. YDS.	A x V
D	S_5	D-1	30	17c	510			
		D-2	50	29c	1450	24.5c	444	10,878
			80		1960			
D	S_6	D-2	50	29c	1450			
		D-3	40	25c	1000	27.1c	500	13,500
			90		2450			
D	S_7	D-3	40	25c	1000			
		D-4	20	10c	200	20.0c	250	5,000
			60		1200			

AREA LINE D = 1,194

SUM A x V = 29,378

Average value of line D = $\dfrac{29,378}{1,194}$ = 24.6c per cubic yard.

[1] From Step 2.

Step 4—BLOCK VOLUME AND VALUE

$$\text{BLOCK VOLUME} = \frac{\text{Area of line C + area of line D}}{2} \times \text{distance between lines}$$

$$= \frac{2,077 \text{ sq. yds.} + 1,194 \text{ sq. yds.}}{2} \times 250 \text{ yds.}$$

= 408,875 cubic yards

$$\text{AVERAGE VALUE} = \frac{\text{Sum of A x V}}{\text{Sum of A}}$$

$$= \frac{34,615 + 29,378}{2,077 + 1,194}$$

= 19.5c per cubic yard

GROSS VALUE OF BLOCK = 408,875 cubic yards x $0.195

= $79,730

59

Bulk samples are sometimes used to check the reliability of other types. In placer drilling, for example, shafts may be put down over selected drill holes to establish a correction or "shaft" factor which when combined with experience-based judgement can be used as a basis for adjusting the calculated drill values as a whole.

h. *Grab samples*: One thing to avoid or to use with caution when investigating placers is the so-called grab sample. This method of sampling relies on a theory that if enough small "grabs" or portions of material are *impartially* taken and then combined into a single sample, the combined material will be representative of the deposit or of the exposure, as the case may be. But in practice the average person fails to make allowance for the large rock or boulders normally found in bank-run material and, as a result, their grab samples contain a deceptive proportion of fines. Such oversight can cause serious error in the estimated value, particularly when based on a few pans of material.

Every mineral examiner will at one time or another run into a situation where systematic sampling is not practicable or is not called for. For example, a property may have no discernible mineralization or mineral-bearing structure, or the mineral examiner may for other reasons be sure that no significant values exist. Here, a few check samples can serve as "written proof" and as such carry far more weight than a bald statement of fact. Grab samples, if fairly taken, serve this purpose well. Grab samples can sometimes be used to check the presence or the distribution of gold in a gravel exposure and thus used as a guide for subsequent sampling and, judiciously used, they provide an excellent check against salting.

Other uses for grab samples could be cited but the novice will be better served by stressing the fact that in nonexpert hands they are almost sure to be misleading. With few exceptions, grab sampling cannot be considered a valid method for testing placers.

i. *Drift mine sampling:* Underground mining procedures are collectively known as "drift mining" where applied to placer gravels. Placers buried under deep masses of low-value gravel or capped by lava flows have been extensively worked in this manner.

Drift mining presupposes the concentration of values in a well defined stratum or channel. Pay streaks in these channels are typically confined within narrow limits, both laterally and vertically, and are commonly related to troughs or other depressions in the bedrock. But it is not unusual to find several generations of "pay leads", as the miners call them, within a given deposit where several periods of erosion and deposition have superimposed successive channels over one another. Such complications are usually quite difficult to decipher and speaking generally, drift-mine deposits seldom lend themselves to the usual investigative procedures.

These deposits are usually explored by adits or shafts from which drifts are driven into the richer portion of the gravels. Crosscuts are run from the main drifts to establish the outer pay limits and at the same time to block the ground for mining. In the usual drift mining venture, exploration, sampling, development and mining are of necessity integral parts of an overall package.

this is particularly true where the values are confined to sinuous, deeply buried pay streaks. In deep ground the cost of close-interval drilling can be expected to far outweigh its benefits. Drilling can, however, be of great value in guiding an underground exploration program when dealing with this type of placer. A relatively few well-placed drill holes may indicate the course of the pay channel, depth and character of the gravel, grade of bedrock, etc. This type of information is essential to a successful development program where drain tunnels and haulage levels are to be kept below the lowest point of the pay channel. An application of this type of churn drilling at the Vallecito Western drift mine has been described by Steffa (1932, pp. 4-6).

When evaluating drift mines, prospect data are usually meager and the mineral examiner should take into account the fact that drift mining, by its very nature, is typically a high-risk venture. He should also recognize that sampling, in the usual sense, may not be practicable.

j. *Hydraulic mine sampling*: Determining the mineral value of large deposits of the type worked by hydraulic mining is perhaps the most difficult task faced by today's mineral examiner. A typical large hydraulic deposit presents most of the sampling problems encountered in drift mines plus a few of its own. Anyone familiar with the immense masses of Tertiary gravels that have been partially worked in California's Sierra Nevada region need not be told that detailed sampling of such deposits ahead of mining would be too costly for practical application.

The California hydraulic miner was no stranger to mining risks and, for the most part, he concurred with Whitney (1880, p. 369), who in assessing hydraulic mine sampling said: ". . . the amount of gold in the gravel cannot be accurately determined by assay or any other proper method. All that can be positively known is the amount obtained in the clean-up." Whitney went on to say: "The miner, by panning a sufficient number of samples, judiciously selected, can form a pretty good idea whether the gravel is likely to pay for working; but this is not by any means the same as ascertaining the exact amount of gold which it contains."

As indicated above many of the early hydraulic operators relied more on qualified estimates of value than on a formal sampling program. Such estimates were based largely on the production records or estimated gold yield of nearby or similar properties but, in most cases, the question of actual gold value was (and remains) secondary to such factors as the area of minable ground, its character, width and course of the deposit, the depth and slope of bedrock, facilities for washing, dump room, water supply, cost of water, etc.

It is obvious that some factors cannot be readily appraised without some underground exploration. Such work, where provided, is usually in the form of a few shafts and some drifting or small-scale operations. Churn drilling may be employed to a limited extent to investigate the course of the channel or bedrock contours. Some examples follow:

The initial sampling and exploratory work at the Malakoff mine (one of California's largest hydraulic operations) consisted essentially of four shafts and 2,000 feet of drifting along the channel, which, according to Bowie (1885, p. 88), cost a total of $63,956.20. It should be noted that at the time (circa 1880), the expenditure of this great a sum for prospecting and sampling was considered somewhat remarkable.

According to Eassie (1944, p. 93), the initial prospecting and sampling for a medium-size hydraulic property (150-foot banks) in British Columbia was done hydraulically, with exception of a small amount of churn drilling undertaken to determine the course of the channels. In all, six exploratory pits were opened but only one supported a continuing operation.

Existing faces in old hydraulic pits can, in some cases, be sampled by conventional means. At the McGeachin mine in Placer County, California. old hydraulic banks roughly 50 feet in height were sampled by means of vertical hand-dug cuts 2 feet wide by 1 foot deep. The sampling crew worked from long ladders propped against the gravel faces.

A small hydraulic mine on Fox Creek, Alaska, was prospected and simultaneously worked in the following manner: The deposit, consisting of low bench gravels flanking a small creek, was superficially tested by panning and its 15 to 20-foot depth was determined by use of 3-inch, hand-driven drive pipes. Several hand-dug pits and shafts were put down in the more promising areas and those returning the best prospects were expanded into ground sluice cuts. Pay areas exposed by ground sluicing were, in turn, expanded into pits and worked out by hydraulicking. Once a pit was started, the hydraulic mining proceeded from clean-up to clean-up without further sampling. When the ground stopped paying, the pit was expanded in another direction or abandoned. In other words, this was a hand-to-mouth operation (as are many small placer mining ventures) and it illustrates the type and extent of sampling or testing generally relied on for working small hydraulic mines.

In most cases the initial prospect work has been engulfed or otherwise eradicated by subsequent mining and, today in most hydraulic areas, there remains little to see but abandoned pits and their sloughed faces. What does this mean to the mineral examiner? It means that in the absence of prospecting shafts or other openings in the unworked gravels, it is necessary to use what past records are available and a large measure of common sense in making an evaluation.

4. SPECIAL PROBLEMS

a. *Large rock or boulders*: At this point something should be said about large rocks or boulders—nemesis of the placer sampler.

When sampling placers there is a natural tendency to bypass areas of heavy boulders or, at least, to take the finer material from among them without considering the end effect on indicated values. In many cases the effect is the same as salting the sample, a fact which does not seem to be recognized by some mineral examiners.

Consider that the gold particles or other valuable minerals in placers are found among the finer-size material while on the other hand, the large rocks or boulders contain no recoverable values within themselves. For this reason, when the sampler takes too much fine material or fails to make an allowance for valueless boulders, he is likely to report an incorrect, high value. To put it another way, where all bank-run material including boulders will be handled in a mining operation, the prospecting should reflect this fact by including a proportionate amount of large material in the samples.

A direct approach to sampling the type of ground under consideration would be to cut samples large enough to *include* the fair proportion of large-size rock and thus obtain samples which truly represent the mine-run material, boulders and all. But where the mineral examiner must work alone or where a limit is placed on the size of his samples through the use of hand tools or for other reasons, the taking of very large samples may not be practicable. In such cases, he must settle for something less than optimum size, but he can guard against possible error in his end valuation by first carefully evaluating the problem and, second, adjusting his calculated sample values (if necessary) to correspond with those which might be obtained from a suitably large-size sample.

In practice this can often be achieved by visually estimating the boulder content of a piece of ground and, where needed, inserting a correction factor in the end sample calculations.

Consider a case where it was estimated that 10 percent of the bank-run material was too large to be represented by material taken from a 3" x 6" x 4' channel cut. The 55.9 pounds of material removed from this cut contained 2.70 milligrams of gold. The indicated value based on these weights was 13.3 cents per cubic yard.

Now, to include a proportionate amount of boulder material this particular sample would have to be 10 percent heavier. In other words, it would weigh 55.9 pounds plus 10 percent, or a total of 61.5 pounds. Using this new weight and the same 2.70 milligrams of gold, the calculated *bank-run* value becomes 12 cents per cubic yard.

In practice the same end figure can be obtained by simply reducing the uncorrected value by a percentage factor equal to the estimated boulder content. In other words, 13.3 cents minus 10 percent equals 12 cents per cubic yard, *bank-run* value.

63

The sample notes and subsequent data sheets must, of course, clearly show all such adjustments and they should preferably show both the uncorrected and corrected figures.

Before leaving this topic it will be pointed out that while the foregoing correction procedure is generally reliable, there are placers having boulder problems so severe, they cannot be effectively sampled by any means short of mining the entire deposit.

b. *Erratic high values*: The usual methods for calculating the value of placer ground rely on an assumption that the value found in a particular sample extends halfway to the next sample. In effect, this assumes that the mineral content changes at a uniform rate between sample points. While valuation procedures based on this concept work well for ground having a general low or uniform average, the finding of erratic high-value samples can create problems. In practice it has been found that to include them at full value will often result in an over-valuation of the lands involved. There are several ways to approach this problem:

First, the sections in which they occur should preferably be resampled. If the resulting value is low, it is safer to use the low figure and this is sometimes done. Where the two results are about the same, most engineers will use the average.

Another commonly used method for adjusting erratic high values is to decide what the highest value reasonably expected from the particular ground might be. This is estimated after considering the type of deposit being dealt with, the overall sample results and in particular the neighboring samples. This might be called the "normal" high value expected in a given property or locality. The erratic "too-high" value is then downgraded to match this figure.

But where it is reasonable to believe that the property as a whole is made valuable by occurrence of erratic, high-value areas, it may be proper to use all sample results at their full value.

There are, of course, other approaches to the problem but in any event the method for handling erratic high values should be governed by conditions found in the particular deposit being investigated. Needless to say, no method of discounting high values should be adopted until the reason for their occurrence is understood.

5. UNCASED OR SMALL-DIAMETER DRILL HOLES

Various attempts to sample gold placers by use of uncased or small-diameter drill holes have for the most part generated more problems than they have solved. As noted earlier, the volumetric gold-to-gravel ratio in 15-cent ground is about one to a hundred million (1:100,000,000). This tells us that calculated drill hole values are extremely sensitive to stray gold particles or incorrect sample volumes. The effect of one gold particle too many or one particle too few is shown in Table II.

TABLE II. – *Error in calculated value cased by one gold particle too many or one particle too few, in a 1-foot drive. Values shown are based on gold weights determined by the author, and gold at $35 per ounce.*

Diameter of Drill Hole	SIZE OF GOLD PARTICLE		
	20-Mesh (6.57 mg.)	40-Mesh (0.91 mg.)	60-Mesh (0.27 mg.)
7½"	58c /cu.yd.	8c /cu.yd.	2½c /cu.yd.
5¼"	$1.18 "	16c "	5c "
3"	$3.60 "	50c "	14c "

It can be seen that any sloughing of an uncased drill hole, should it occur in a gold-bearing zone, will tend to upgrade the calculated value by significant amounts. If one knew with certainty how much excess material entered a sample and exactly where it came from, the discrepancies might be appraised and allowances made during the calculation process. But experience tells us that this is not always practicable and that when dealing with the usual gold placer, uncased drill holes just do not give reliable results. More often than not the ground will be over-valued and this seems to be particularly true when rotary drills are employed, especially those using roller-type bits. In one case, placer ground reported to run 26 cents per cubic yard on the basis of samples obtained from rotary drill holes was found to contain 1 1/4 cents per cubic yard when checked by an experienced dredging company using a Keystone placer drill and qualified crew.

The mineral examiner should *always* view critically any placer evaluation based on gold values obtained from uncased or small-diameter drill holes. Figure 15 shows why, in the case of small-diameter holes, the indicated values are not always reliable.

While some gold dredging companies occasionally employ 4-inch churn drills for preliminary prospecting, the 6-inch size equipped with a 7 1/2-inch diameter drive shoe is considered standard.

Under certain conditions, small-diameter drill holes may be both practical and adequate. As an example, when evaluating beach-type deposits valuable for their chrome or titanium-bearing minerals it is often possible to rely on small-diameter (say 2-inch) holes for sampling. Here the fine-grained nature of the deposit, the relatively large proportion of valuable mineral and its low unit value combine to eliminate some of the uncertainties normally associated with small-diameter drill holes. As another example, consider a tin placer containing 1/10 lb. cassiterite per cubic yard. With tin at $1.50 per pound, the indicated value would be 11.6 cents per cubic yard and the volumetric ratio of cassiterite to gravel 1:114,000. In other words, the

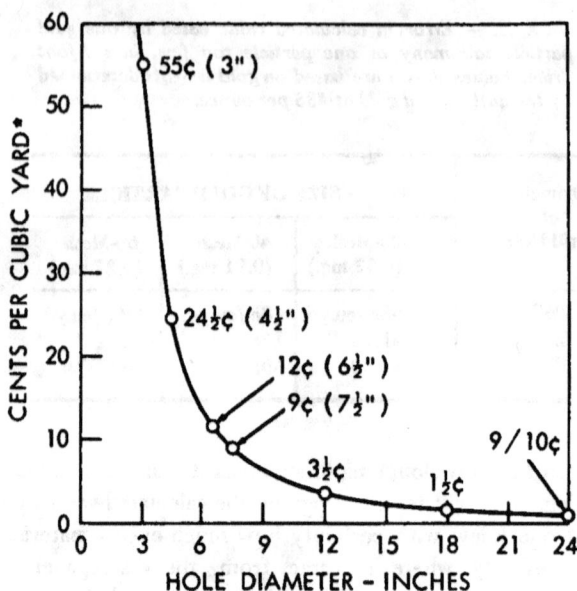

Gold @ $35.00/oz.

FIGURE 15.— *Curve showing change in calculated value caused by plus or minus 1 milligram of gold in a 1-foot sample. It can be seen that accidental loss or addition of a single gold particle can cause serious error when dealing with small-diameter drill holes.*

volume of valuable mineral in this case would be approximately 1,000 times greater than the volume of gold in a comparable gold-placer and this illustrates why low-value placer minerals are, in general, easier to sample.

6. **SALTING**

There are two kinds of salting—intentional and innocent. Intentional salting can be defined as the surreptitious addition of valuable material to a sample with intent to deceive. Innocent salting, which can have the same end effect, can be accidental or the result of carelessness or improper working procedures. Although intentional salting is seldom encountered, only the engineer who is constantly alert can feel secure about his sample results. A few of the salting methods applicable to placers are:

Secreting gold in the material to be sampled.

Placing gold in sample containers or in the sample itself.

Secreting gold in the excavating or washing equipment.

Dropping bits of gold in the pan during final processing.

Where the volume of material obtained from a drill hole or from a channel cut is comparatively small, it takes very little gold to appreciably alter the value. It should be noted that in bulk sampling a substantial amount of gold may be required to achieve the desired effect but, at the same time, the opportunity for salting may be greater. This is because the larger the sample the more it is handled, and the more it is handled, the more difficult it becomes to provide adequate safeguards or keep it under constant surveillance. Ironically, it may be easier to detect salting after it has happened than to prevent it in the first place.

But there are many precautions that can be taken to discourage salting, most of which require little more than common sense. Some are:

Keep the exact location of the point to be sampled a secret until the actual cutting begins.

Clean and carefully inspect all faces before sampling.

Clean and carefully inspect all equipment before sampling.

Check equipment at appropriate intervals.

Complete the cutting and processing during a single work-day if possible.

When sampling is completed, go back and take a few check samples from selected points, particularly those showing unusually high values.

Some placer drillers place barren material in the bottom of the drill hole at the end of the shift and test it for values when resuming operations. The black sand concentrate obtained from a sample should be examined with a hand lens or preferably, with a low-power microscope because at this point

FIGURE 16. — Photomicrograph of gold shavings used to salt a placer sample (X 100). This fraud was detected by microscopic examination of black sand concentrates obtained from the sample.

the careful observer can usually detect any anomolous or foreign material in the sample. Figure 16 is a photomicrograph showing gold shavings found in a placer sample. In this case, the shavings were small enough to be missed in a cursory inspection of the pan concentrate but examination under a 20X binocular microscope revealed the fraud.

Another give-away may be sample results that are consistently too high or too regular in value. The mineral examiner should always compare his sample results with the known or expected tenor of the district and where marked discrepancies appear, the cause should be ascertained.

While the possibility of samples being tampered with must never be disregarded, the possibility of the sampler "salting" himself through carelessness or through use of an incorrect procedure is usually much greater.

7. WHEN SAMPLING FAILS

Not all placers lend themselves to the usual forms of sampling and there are some which cannot be sampled by any means short of mining the entire deposit. This problem usually reflects one of two things: (1) The gold or other mineral occurrence is so erratic that usual sampling or evaluation methods break down hopelessly. (2) There are physical conditions which make sampling impractiable.

Excellent examples of the latter condition can be found in California's drift mines. Sampling problems inherent in this type of deposit have already been brought out under the heading of "Drift Mine Sampling" but the point to be made here is that the physical characteristics of such deposits when combined with an almost complete absence of surface guides to their underground location makes sampling in the normal sense impracticable. Historically, the successful exploration and development of this type of deposit has relied on the practical application of bits and pieces of information gleaned from limited surface exposures or underground workings. This is another way of saying the old-time drift miner relied largely on geologic inference tempered by good judgement. Today's mineral examiner, when confronted by a deposit that cannot be readily sampled, may also have to fall back on his powers of observation and other resources.

Where factual sample data cannot be obtained without extensive and costly explorations, which except in isolated cases will probably never be made, authentic production records will generally be found a useful source of information. And even where the actual records are not available, it may be possible to obtain enough statistical information from state mining bureau reports or other government publications to ascertain the past production of better-known mining districts and to appraise their remaining potential. To varying degrees, this type of statistical data can be applied to individual properties. Noteworthy examples are found in studies of California's hydraulic mine reserves (Jarman, 1927) carried out by State and Federal agencies, preparatory to construction of the North Fork and Yuba Narrows debris retention dams.

But, the point to be stressed is that sooner or later the mineral examiner can expect to encounter a placer deposit where sampling fails and when this happens, it is important that the fact be recognized.

REFERENCES CITED (PART III)

Bowie, A.J., 1885, A practical treatise on hydraulic mining in California: D. Van Norstrand Co., New York.

Daily, Arthur, 1962, Valuation of large gold-bearing placers:
Engineering and Mining Journal Vol. 163, No. 7, July, 1962, pp. 80-88.

Doheny, L. C., 1941, Placer valuation in Alaska, Part I:
Engineering and Mining Journal, Vol. 142, No. 12, Dec., 1941, pp. 47-49.

Doheny, L. C., 1942, Placer valuation in Alaska, Parts II and III:
Engineering and Mining Journal Vol. 143, No. 1, Jan., 1942, pp. 43, 44 and Vol. 143, No. 3, March, 1942, pp. 48, 49.

Draper, Marshall D., 1932, Placer ground sampled by well-digging equipment:
Engineering and Mining Journal Vol. 133, No. 10, Oct., 1932, p. 537.

Eassie, W.H., 1944, What does hydraulicking cost? Engineering and Mining Journal, Vol. 145, No. 6, June, 1944, pp. 92-94.

Harding, James E., 1952, A new way to interpret the results of placer sampling:
Engineering and Mining Journal Vol. 153, No. 1, Jan., 1952, pp. 96-98.

Herr, Irving, 1916, Sampling placer-gravel deposits: Engineering and Mining Journal, Vol. 102, No. 6, Aug. 5, 1916, pp. 261, 262.

Huttl, John, 1941, Unique portable sampler for shallow placers:
Engineering and Mining Journal Vol. 142, No. 9, Sept., 1941, pp. 55,56.

Jarman, Arthur, 1927, Report of the Hydraulic Mining Commission upon the feasibility of the resumption of hydraulic mining in California (a report to the California Legislature of 1927): Twenty-third Report of the State Mineralogist, Jan., 1927, pp. 44-116.

Also see-Jadwin, Edgar, Hydraulic mining investigations in California (a report of the War Department, Board of Engineers for Rivers and Harbors, dated April 20, 1928):
Senate Doc. No. 90, 70th Cong., 1st Sess.
Also see—Bradley, Walter W., Dams for hydraulic mining debris:
Calif. Jour. of Mines and Geol., Vol. 31, No. 3, July, 1935, pp. 345-367.

Lorain, S.H., 1944, Prospect trenching with caterpillar-mounted angledozers:
U.S. Bureau of Mines Information Circular 7294 (Sept., 1944), 8 pp.

Magee, James F., 1937, A successful dragline dredge: Amer. Inst. of Min. and Met. Engr., Trans. Vol. 126 (1937).

Prommel, H.W.C., 1937, Sampling and testing a gold-scheelite placer deposit in the Mojave Desert, Kern and San Bernardino Counties, California: U.S. Bureau of Mines Information Circular 6960 (1937), 18 pp.

Sawyer, Dwight L., 1932, Sampling a gold placer: Engineering and Mining Journal Vol. 133, No. 7, July, 1932, pp. 381-383.

Steel, Donald, 1915, Prospecting wet placer ground by shaft sinking: Mining and Scientific Press, Vol. 110, No. 2, Jan. 9, 1915, pp. 66-68.

Steffa, Don, 1932, Gold mining and milling costs at the Vallecito Western drift mine, Angles Camp, California: U.S. Bureau of Mines Information Circular 6612 (1932), 14 pp.

Whitney, J.D., 1880, The auriferous gravels of the Sierra Nevada of California: Contributions to American Geology, Vol. 1; Harvard University Press, Cambridge, Mass.

Woodbridge, T.R., 1916, Ore-sampling conditions in the West: U.S. Bureau of Mines Tech. Paper 86 (1916).

PART IV

SAMPLE WASHING EQUIPMENT

1. GENERAL CONSIDERATIONS

Equipment used for the recovery of placer gold has changed very little over the years and, in general, remains relatively simple. Most devices employ some form of riffled surface to hold the gold or other heavy mineral after it has been separated from the valueless material. The actual separation relies on the ability of heavy minerals to resist the action of moving water while the lighter materials are carried away. In dry washers where a current of air is used as the transporting medium, the same principle applies. Although many have tried, no one to date has devised a gold-saving device or system which can economically replace the ordinary riffled sluices or placer jigs used on today's dredges or in other comparable large-scale placer operations. It is true that sluices may lose some of the fine gold but his is normally offset by their low operating cost and their high unit capacity which combine to return the greatest dollar profit.

When selecting a machine for washing and concentrating placer samples, the first consideration should be whether or not it will indicate the *commercially recoverable* gold content of the sample. Other desirable features would be:

Low first cost.

Easy to maintain or repair in the field.

Easy to transport and set up for operation.

Will accept bank-run material without pre-screening.

Will thoroughly wash and reject the oversize material.

Makes efficient use of water.

Will efficiently process small as well as large samples.

Will effectively reduce the sample to a small-volume residue or concentrate.

Can be quickly and completely cleaned between samples.

Is time-tested and accepted by knowledgeable mine operators or engineers familiar with placer sampling requirements.

It should again be stressed that no dredge or other large-scale placer equipment saves 100 percent of the values and because of this it is important that the sample washing process indicate the actual returns to be expected from a commercial operation. In this connection it is noteworthy that the pan, rocker, and the sluice when used by experienced placer operators fulfill this requirement.

2. MINERS' PAN

The miners' pan, better known as a gold pan, is perhaps the most widely used device for washing small placer samples and from the standpoint of simplicity it has no peer. A separate article describing its use and manipulation will be found in Part V under the heading of "PANNING." In the hands of an expert the pan is both a versatile and a highly efficient concentrator well suited to washing small amounts of gravel but where individual samples weigh more than about 30 pounds, or where a large number are to be washed, something which provides a greater throughput with less expenditure of time and labor is needed.

3. SLUICE BOX

Another widely used sample washing device is the ordinary sluice box which in its smaller form is sometimes loosely but erroneously called a "long tom". A sluice in its simplest form is no more than an elongated, rectangular trough fitted with transverse cleats or some other form of riffled bottom. It is operated by essentially allowing a stream of water to carry the sands and gravel over the riffles which, in turn, detain any gold or heavy minerals as they settle to the bottom. Small sluices of the type used for sampling are commonly 8 or 12 inches wide with 6 or 8-inch sides and are usually 6 to 12 feet in length. Construction details and the materials used are largely a matter of personal choice, but the simpler ones are no more than an open-ended trough made of planed 1-inch lumber and provided with cross bracing where needed. Some samplers prefer a sluice made of heavy-gauge sheet metal (made rigid by bracing) and others prefer exterior-grade plywood painted with marine varnish or spar. The longer sluices are usually sectionalized to facilitate transportation.

The primary function of a riffle is to retard heavy minerals giving them a chance to settle and, at the same time, to provide pockets in which the values are retained. A well designed and properly working riffle will create eddies along its downstream edges and the so-called "boiling" action in these eddies will do two things. First, it will prevent packing of the black sand or other material caught by the riffle, and, second, it will provide a classifying effect which further concentrates the gold and heavy minerals. While this boiling action must be strong enough to prevent packing it must not be so strong that flake or fine gold cannot settle. It should be obvious that proper riffle action is the key to good recovery. An easily built and effective riffle suitable for use in a small sampling sluice can be made from ½" x ½" wood strips placed across the bottom of the sluice at right angles to the flow, with 3/4" spaces between each riffle bar. The boiling action can be improved by undercutting the downstream face of each bar on a 30-degree bevel. The individual riffle bars are tacked to wooden side rails and the whole assembly held in place by means of cleats and wedges as shown in Figure 17. The riffle assemblies are made a littler narrower than the inside width of the sluice and of convenient length.

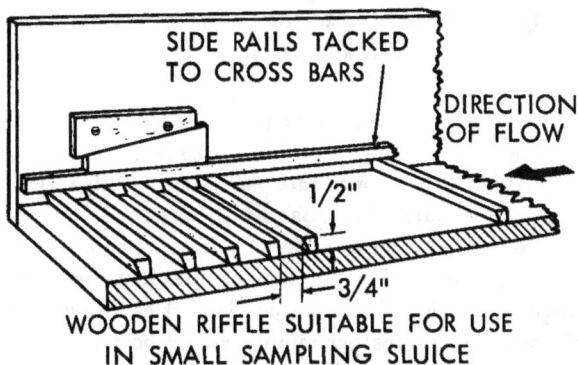

SIDE RAILS TACKED
TO CROSS BARS

DIRECTION
OF FLOW

1/2"

3/4"

WOODEN RIFFLE SUITABLE FOR USE
IN SMALL SAMPLING SLUICE

FIGURE 17. — *Wooden riffle suitable for use in a sampling sluice.*

Heavy wire screen of the type used for screening gravel, and expanded metal lath are sometimes used as riffles in small-size sluices, particularly where much fine gold is present. This type of riffle is usually installed over burlap mats, cocoa matting, or similar rough-surfaced fabric which helps hold the fine gold. Because burlap and cocoa matting are difficult to clean, ordinary canvas should be substituted in sampling sluices. It will be noted that metal webs forming the diamond-shaped openings of expanded metal lath have a noticeable slope in one direction. When the expanded metal is placed in the sluice so this slope leans downstream, small eddies form beneath the over-hangs and make excellent gold catchers. Expanded metal riffles do not have a large holding capacity, that is, they may soon fill with concentrate, but this is seldom a problem in sampling work where close watch and frequent clean-ups should keep the riffles working efficiently. Hungarian-type riffles, such as those shown in Figure 17, can hold a comparatively large amount of concentrate and for this reason may be preferred where the gravel contains much black sand. On the other hand, expanded metal riffles leave a minimum amount of concentrate to be panned at the end of a sample run and this time-saving feature makes them easy to clean up. Many engineers compromise by equipping the upper 2 or 3 feet of a sampling sluice with Hungarian-type riffles and the remainder with expanded metal lath.

In commercial-scale placer operations, mercury is usually placed in the riffles to assist holding the gold over extended periods of time but in small-scale sampling work where clean-ups are frequent, mercury is not needed for this purpose and is seldom used.

It should be noted that when a particle of gold is "wetted" by mercury, the mercury actually penetrates the surface and causes the gold to become brittle. Depending on the size of the gold particle and length of exposure, the penetration may be superficial or complete. The ductility is not restored when the mercury is removed with acid and in the case of small gold particles, a delicate crystalline structure is often induced. It can be seen that amalgamation within the sampling sluice would either impair or prevent a

75

later study of particle size, the surface texture, or other physical characteristics of the gold, any of which could prove important in a placer investigation.

The feed and water flow should be regulated to maintain proper riffle action and it can be said generally that where this is not done more fine gold will be lost by its inability to penetrate packed riffles, than will be carried off by suspension in the water. The quantity of water required varies considerably according to the character of material being washed and the rate of feed, the type of riffle, width and grade of sluice, etc. Because of these variables the water requirement is difficult to predict but as a rough guide, a minimum flow of about 50 gallons per minute should be provided for a 6 or 8-inch sampling sluice. Unlike panning or rocking, the water is seldom reclaimed or reused and, for this reason, the water requirement for a typical sampling operation can easily be several thousand gallons per day. It should be apparent that the sluice is best suited to testing stream placers. Supplemental equipment is often needed. This may be a water pump, a puddling box or tub, or some kind of screening arrangement. Because of the relative short length of the usual sampling sluice, cemented gravel or gravel containing much clay must be thoroughly broken up in a puddling box or tub before being fed into the sluice. Screening out the plus ½-inch rocks ahead of the sluice will conserve water and will generally improve the entire operation.

The slope or grade of a sluice depends on a number of factors including rock size and shape, the amount and composition of sand, type of riffle and depth of water flow. In each case the proper grade must be determined by trial and this can be best done by initially setting the sluice on a grade of about 1 inch fall per foot of length and then adjusting the grade as found necessary. When in doubt it is better to have the grade too steep than too flat, as too flat a grade will result in sanding of the riffles which in turn will impair their ability to recover fine gold. Where fine gold is to be saved the depth of flow over the riffles should be as shallow as possible while still sufficient to carry off the pebbles and maintain a loose bed between the riffles. To do this the sluice grade is steepened and it can be said generally that the grade for very fine gold should be steeper than for coarser colors. Increasing the grade will, to a point, offset a deficient water supply but, in any case, the riffle action should tell the operator when the proper balance has been reached.

The daily capacity for a sampling sluice varies widely with the type of gravel, degree of cementation, amount of clay, etc. These factors individually and collectively determine the amount of material that a man can prepare for washing in a day's time and, in many cases, the preparation time will exceed the actual washing time. Under favorable conditions with an efficient sluice set-up, two men can handle 1 to 3 cubic yards per day taking into account time needed for clean-ups between samples, processing the sluice concentrate, logging sample data, etc.

When the mineral examiner washes a placer sample in a small sluice box and fails to find the amount of gold anticipated by the property owner or vendor, it is sometimes argued that he failed to recover the hoped-for value because his sluice was too small or the sample was put through too fast. While it is true that a sluice box crowded beyond its optimum capacity will lose some of the gold, a quick look at the facts will usually show that there is little or no room for argument in most cases.

For example, modern gold dredges provide about 10 square feet of riffle area for each hourly yard of material passing over the gold-saving tables. By direct comparison it can be shown that it would be necessary to feed an 8-inch x 10-foot sampling sluice at a rate of 1 1/2 cubic yards per hour to attain this degree of riffle loading. But experience tells us that the rate of feed for a sampling sluice of this size is more likely to be on the order of 1½ cubic yards *per day* rather than per hour. It can be seen that in a sampling sluice the riffle area provided for each unit of sample material is considerably greater than the riffle area provided in standard mining practice. In other words, the small sampling sluice usually favors the sample. This is borne out by long experience which shows that a properly constructed and carefully operated sampling sluice will save all of the gold or other heavy minerals which can be economically recovered by standard placer methods.

4. ROCKERS

Rockers are widely used for washing placer samples and under most conditions they are well suited to the needs of the field engineer. It should be noted that since its inception, the American gold dredging industry has, for the most part, used rockers in preference to other types of sample washing equipment. The fact that the rocker has survived in direct competition with a variety of more "modern" or "improved" machines which have been introduced over the years attests to the reliability of results which can be obtained.

Rockers are usually homemade and are built in a variety of sizes and designs depending on the ideas and experience of the builder. Figure 18 shows a lightweight rocker which is easily built and is suitable for general use. Individual components are a screen-bottomed hopper for receiving and sizing the feed material, a canvas apron for deflecting the feed to the head-end of the rocker, the rocker body which is equivalent to a sluice, and a bed plate or frame upon which the entire assembly is mounted. A rocker may be made any convenient size because within limits, its size will determine throughput capacity rather than its efficiency as a gold saver. Although there are no fixed size criteria, a rocker 1 foot wide by 3 feet long would generally be referred to as "small" and one more than 18 inches wide or 5 feet long as "large." When made of wood (which is usually the case) clear lumber without cracks or flaws must be used and the bottom should be made of vertical-grained stock which will not shred or rough-up when

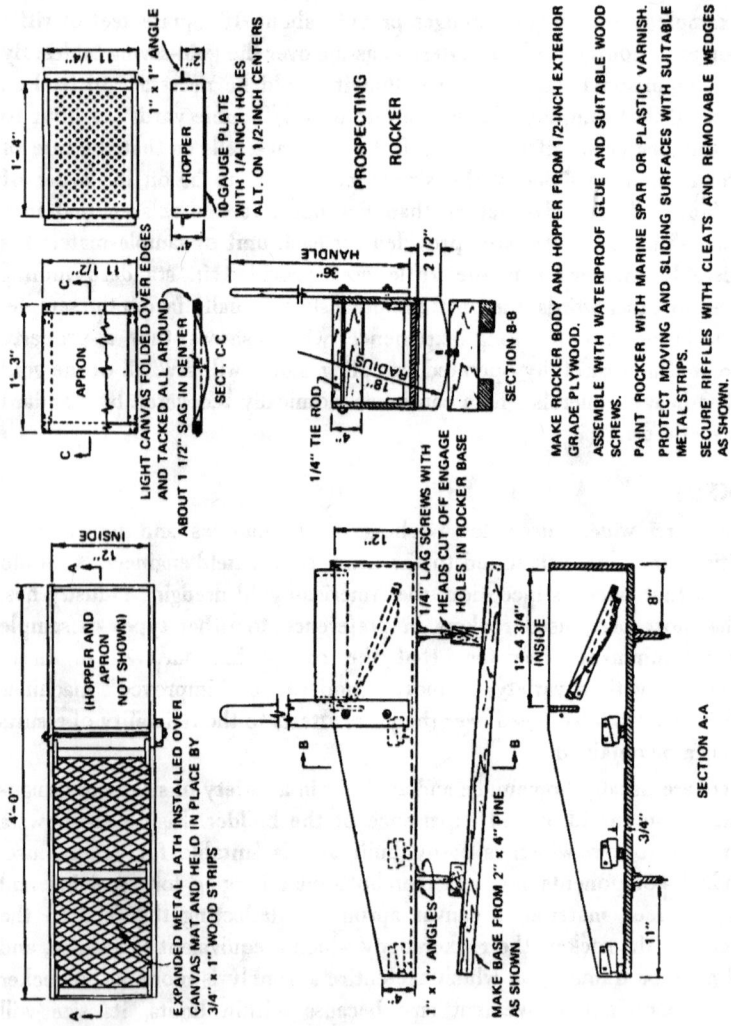

11 1/4"

1-4"

1" x 1" ANGLE

1 1/2"

HOPPER

10-GAUGE PLATE
WITH 1/4-INCH HOLES
ALT. ON 1/2-INCH CENTERS

PROSPECTING

ROCKER

11 1/2"

1-3"

APRON

C

C

LIGHT CANVAS FOLDED OVER EDGES
AND TACKED ALL AROUND
ABOUT 1 1/2" SAG IN CENTER

SECT. C-C

HANDLE

36"

2 1/2"

18" RADIUS

4"

SECTION B-B

1/4" TIE ROD

MAKE ROCKER BODY AND HOPPER FROM 1/2-INCH EXTERIOR
GRADE PLYWOOD.

ASSEMBLE WITH WATERPROOF GLUE AND SUITABLE WOOD
SCREWS.

PAINT ROCKER WITH MARINE SPAR OR PLASTIC VARNISH.
PROTECT MOVING AND SLIDING SURFACES WITH SUITABLE
METAL STRIPS.

SECURE RIFFLES WITH CLEATS AND REMOVABLE WEDGES
AS SHOWN.

INSIDE

12"

A

(HOPPER AND
APRON
NOT SHOWN)

4-0"

A

EXPANDED METAL LATH INSTALLED OVER
CANVAS MAT AND HELD IN PLACE BY
3/4" x 1" WOOD STRIPS

12"

1/4" LAG SCREWS WITH
HEADS CUT OFF ENGAGE
HOLES IN ROCKER BASE

B

B

1" x 1" ANGLES

MAKE BASE FROM 2" x 4" PINE
AS SHOWN

1-4 3/4"

INSIDE

8"

3/4"

11"

SECTION A-A

FIGURE 18. – Prospecting rocker suitable for general use. With this rocker, one man working under normal
conditions can wash the equivalent of eight 80 pound samples per day. This includes the time needed for
clean-ups, panning, and logging sample data.

78

scraped. Heavy-duty rockers and those which are in continuous use are preferably made of sugar pine reinforced with suitable metal fittings. Where sugar pine is not available, redwood makes a fair substitute. Rockers made of exterior grade plywood have the advantage of lighter weight and they are strong and durable if the joints are sealed with waterproof glue. The exposed surfaces of a plywood rocker should be protected with marine varnish or spar.

The hopper should be made 1/2 to 3/4-inch narrower than the inside width of the rocker; this results in a bumping action as the rocker is operated and assists scouring and screening the gravel. The amount of open area in the screen plate is important. A screen with too much open area allows the fines to pass through too fast and where much sand is present this can cause overloading of the riffles with resultant gold loss. A plate with 1/4-inch holes drilled or punched in an alternate pattern at 1-inch intervals and on 1/2-inch centers will provide a screen with 10 percent open area. This has been found to be about right for small prospecting rockers.

The apron consists of a simple wooden frame covered with loose-fitting canvas or similar material. The resultant sag or "belly" in the apron functions as a gold and black sand trap and when large amounts of material are put through a rocker, the apron's gold-holding ability permits longer runs between clean-ups. In normal sampling work where short runs are made, the apron is less important and is sometimes dispensed with.

It is noteworthy that most old-time prospectors employed by dredging companies wash their samples in rockers *without riffles*. Experience and practice enable these men to concentrate the gold and black sand directly on the smooth wooden bottom of the rocker and, as a final step, to tail out the black sand and bring the gold to a point in a manner somewhat similar to panning. This technique is particularly useful when applied to small-volume churn drill samples and is mentioned here to point out that a properly operated rocker is an excellent gold saver which unlike the sluice, does not rely on riffles for its effectiveness. But it should be pointed out that a rocker equipped with good riffles will forgive much mishandling and, for this reason, the average person should use them. Parenthetically, it should be noted that even the experienced rocker operator, in certain cases, may find it more expedient to use riffles than to explain to a layman why they are not essential. A simple riffle arrangement suitable for general use can be provided by covering the rocker bottom with heavy-gage expanded metal lath placed over a canvas mat and held down by several transverse wooden slats as shown in Figure 18.

The length of the rocker handle is important—it should be waist-high to the operator in a standing position. The long leverage thus provided makes the rocker much easier to handle and materially reduces the physical effort which, at best, is considerable. Anyone who has attempted to operate a short-handled rocker from a sitting position will appreciate the foregoing comment.

Like panning, rocking relies to some degree on subtle techniques which must be learned by experience but the following pointers may help the novice get off to a good start. The first step is to set the bed plate and secure it so that it will not shift or move around when the rocker is in operation. The best slope for the rocker will have to be determined by trial but if it is initially set at 1 1/2 inches fall per foot of length, a few short trial runs will suffice to make any needed adjustment. Insufficient grade may cause sand to blanket the riffles and result in loss of fine gold. If the hopper is filled too full, gravel will slop over the sides when rocked and it will be difficult to regulate the flow of material through the screen and, for this reason, the hopper should not be filled over half full and preferably the screen plate should be left partially exposed at one end. Starting at the exposed end of the screen plate, water is poured over the gravel while the rocker is shaken vigorously and the amount of material fed to the riffles is regulated by shifting the point of water application back and forth between the gravel and the exposed screen plate. Most pictures illustrating the use of a rocker show the water being applied with a long-handled dipper such as a gallon can on the end of a stick. In practice one finds that it takes considerable dexterity to use a dipper and at the same time operate the rocker smoothly and maintain a uniform flow of material over the riffles. For this reason, a water hose supplied by a pump or a gravity-flow should be provided where possible. The flow of water obtainable from an ordinary garden hose (about 5 gallons per minute) is usually enough for operating a rocker but where this is not available, two or three barrels of water used in a closed circuit will generally be sufficient for a day's work. When water must be dipped, the water barrel should be placed next to the head-end of the rocker within easy reach of a short-handled dipper.

When rocking is periodically stopped to discharge rocks from the hopper, the material passing over the riffles will settle and tend to pack, particularly if much black sand is present. If washing is resumed without first loosening this material, any fine gold that has not penetrated the packed sand will be lost with the tailings. To guard against this, the material behind each transverse riffle should be loosened and scraped back toward the head-end of the rocker with a flat, straight-edged scoop. *This loosening procedure is the key to effective rocking.* The loosened material is subsequently washed down and re-concentrated with the next run and if the rocker is kept free of packed sand in this manner, it will effectively recover flour gold particles running 1,000 or more to the cent.

After the entire sample has been washed, the concentrate remaining behind the transverse riffles is picked up with the scoop and placed in the upper end of the rocker and then carefully re-washed once or twice with clear water to remove surplus sand and further reduce the concentrate volume. The apron, riffles and canvas mat are then removed and washed out in a pan or tub of water and the combined concentrate panned to a finished product.

Although reference books state that from 1 to 3 cubic yards can be washed in a rocker per man-shift, it should be noted that these figures apply to production-type work and give little indication of what should be expected when processing samples. Experience has shown that one man using a 12-inch x 4-foot rocker similar to the one illustrated in Figure 18 can wash the equivalent of eight 80-pound samples per day where conditions are favorable. This includes the time for clean-ups, panning, and logging sample data. The total amount will, of course, vary with the size or number of individual samples and with the character of material being washed. Material containing much clay can halve this figure whereas loose, free-wash type material may double it. Dredging companies occasionally use large, engine-powered rockers in which the total gravel excavated from a shaft can be washed rapidly and economically. A typical power-driven rocker having a capacity of 5 cubic yards or more per 8-hour shift is illustrated and described on pages 334 and 335 of the August 25, 1923, issue of Engineering and Mining Journal-Press.

5. SPECIAL MACHINES

A variety of small mechanical gold washers have been manufactured and put on the market over the years and although most were intended to be used for small-scale mining operations, some were advertised and sold as prospecting or sampling units. The typical machine consists of a small trommel screen with a feed hopper for shoveling in, a short sluice (which is usually provided with a shaking motion or some kind of "special" riffle), a pump, and a water distribution system, all run by a small gasoline engine. The typical machine weighs more than 500 pounds and requires a pickup truck or a trailer for transportation. Most are no longer manufactured and have passed from the scene but two of the better-known machins are still on the market.

These are manufactured by the Denver Equipment Company and known as the "Denver Mechanical Gold Pan" and the "Denver Gold Saver" respectively. The mechanical pan has been well received in the industry over a 30-year period and is generally referred to simply as a "Denver Pan." It comprises an assembly of three shallow, nested pans 2 feet in diamteter with two superimposed screens arranged to wash and reject the plus ¼-inch material. The combined assembly is mounted on a horizontally gyrating base driven by a small gasoline engine and the resultant motion is said to duplicate hand panning. The minus ¼-inch material after passing through the screens, progressively flows over the three pans, one discharging into the next. The uppermost pan is provided with an amalgamating plate, and the two successive pans with special rubber mats or cocoa matting held down by coarse wire screen. Capacity of a single unit ranges up to 2 cubic yards per hour and water consumption is said to be as little as 1 to 2 parts (by weight) per part of gravel which would indicate an average of about 1,000 gallons per cubic yard. Single or double units are available and these can be provided with a scrubber and a trommel-type screen. The largest (Duplex) unit when so equipped has a rated capacity of 4 to 6 bank-run yards per hour and weighs 2100 pounds. The single (Simplex) unit without the trommel weighs 675 pounds. The Denver Mechanical Gold Pan is sturdily built and is suitable for continuous use such as would be encountered in a mining operation.

The Denver Equipment Company's second machine, sold under the name of "Denver Gold Saver", is well suited for general sampling work in that it can be quickly and easily cleaned thus eliminating the danger of carry-over of values from one sample to the next. It consists of a feed hopper, a combined scrubber and trommel to wash and screen out the plus ¼-inch material, a special vibrating molded riffle, a reserve water storage tank and a centrifugal pump with suitable piping sprays, etc., all powered by a 1½ HP gasoline engine. The complete assembly weighs 750 pounds and has a rated capacity of 2 to 3 cubic yards per hour.

6. DRY WASHERS

In arid districts where water is scarce or expensive and a "dry" plant is proposed for the recovery of placer gold, a small dry washer may be the logical choice for processing samples. A number of small, hand-powered machines are on the market and most work quite well within certain limits common to all dry washers.

First, consider two identical gold particles, one in a dry washer and the other submerged in water. It will be shown that the relative weight of the dry gold particle is substantially less than its wet counterpart. This is illustrated by the following equations in which the specific gravity of gold is 19, the specific gravity of gravel 2.65, and the specific gravity of water is 1:

$$\frac{\text{Gold in air, } 19}{\text{Gravel in air, } 2.65} = \frac{7}{1}$$

whereas

$$\frac{\text{Gold in water, } 19\text{-}1}{\text{Gravel in water, } 2.65\text{-}1} = \frac{11}{1}$$

This shows the gold to be eleven times heavier than gravel when immersed in water, as compared to 7:1 in air. Simply stated — the relative weight of gold is about 1½ times greater when passing through a wet process than when passing through a dry washer. The lesser ratio in air is not too apparent when processing coarse or "shotty" gold but it no doubt contributes to the dry washer's often poor recovery of fine or flaky gold.

To respond to dry washing the material treated must be almost completely dry. In most cases 3 percent moisture is considered a maximum. The gold particles must also be completely liberated and free of cementing material such as caliche. In addition, effective separation is usually dependent on use of a closely-sized feed, sometimes not larger than ¼" screen-size. One might say that the amenability of a dry placer to working by the usual (dry) methods varies in proportion to its suitability to sizing by screening.

Besides the well-known difficulties related to drying placer material prior to dry washing and using small-mesh screens, there are other problems less generally appreciated. Wilson and Fansett (1961, p. 108) in referring to tests made at the University of Arizona sum these up by saying:

> "A dry concentrator will not make as high recovery as a wet concentrator. Under favorable conditions, the recovery will be approximately ten to fifteen percent less with a dry machine as compared with a wet machine."

This tells us that in cases where a dry processing plant is proposed for the recovery of gold in a commercial-scale placer operation, it might be advisable to process the initial samples with a small dry washer. Figure 19 illustrates a type of machine found suitable for this purpose. If all else is equal, the recovery obtained from the dry (sample) washer should indicate the recovery to be expected from the full-scale operation. On the other hand, if the samples are washed in a rocker or other wet device, a suitable recovery "correction" factor may have to be applied to the indicated sample values.

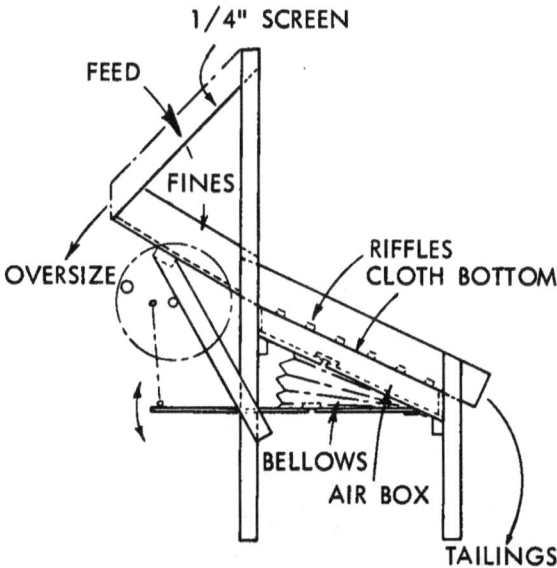

FIGURE 19 — Schematic diagram showing the arrangement of a typical hand-operated dry washer. In use, the dirt is moved over the riffles by puffs of air forced through the cloth bottom by the bellows. The gold lags behind and accumulates along the upstream edges of the riffles.

REFERENCES CITED (PART IV)

Wilson, Eldred D. and Fansett, George R., 1961, Gold placers and placering in Arizona: Arizona Bureau of Mines Bulletin 168 (1961).

PART V

PANNING AND
ASSAY PROCEDURES

1. PANNING

Panning is a simple operation but, at the same time, it is difficult to describe. Although the subtle techniques of panning vary with the individual and with the material being washed, the overall operation can be divided into six basic steps as follows:

a. *Preparation:* After filling the pan approximately level full, carefully submerge it in quite water, preferably resting it on the bottom of a shallow pool or tub with the top of the pan just below the water surface. After the material has become thoroughly wet, work over the contents with both hands and break up any lumps. If clay is present, knead and stir the material until the clay is dissolved and floated away. It is important that all clay be eliminated before actual panning begins. Wash off and throw out all large rocks. In this first step the eye and hands substitute for a screen.

b. *Suspension and stratification:* Commence this stage by grasping the pan with hands on opposite sides and while holding the submerged pan level, twist it back and forth (clockwise and counterclockwise) with sufficient vigor to keep the contents loose. This allows the heavier minerals to migrate to the bottom of the pan and is similar to the action in a jig in which heavy mineral grains are separated from lighter grains by their ability to sink through a semi-fluid bed. If this second step is properly executed, the smallest and heaviest grains will migrate toward the bottom and the larger and lightest to the top. This will allow many of the pebble-size rocks to be manually removed by raking them out of the pan with the fingers.

c. *Washing:* The third step is one which, depending on the nature of the material being washed, may take on many variations. It is like Step 2 in that the entire contents of the pan are kept in motion, but as stratification of the bed develops, the lighter particles are allowed to escape over the rim of the pan. To do this, raise the pan partially above water and move the hands slightly back of center (allowing the pan to tip forward with the low side away from the panner) and change from the twisting motion of Step 2 to a flat circular motion. While keeping the pan partially submerged and its contents loose, gradually work the lighter-weight material over the low side of the pan. The rate of discharge is regulated by raising or lowering the pan rim and by using a side to side motion along with the flat circular motion. Alternate Steps 2 and 3 and wash until the bed begins to pack or until heavy minerals begin to crowd to the surface.

d.   *Cleaning:* The fourth step involves selectively washing away surface grains and, in effect, it can be compared to the action of wash water on a concentrating table. To prepare the now partially concentrated material for this step, the pan is given a short, quick side to side motion of sufficient vigor to thoroughly loosen the bed and further stratify the material. During this shaking phase the pan is tipped gradually forward until the surface of the mineral bed becomes flush with the lip. At this point the shaking is stopped and the mineral bed allowed to settle. Next, a thin layer of the lighter material is removed by carefully dipping and raising the pan with a forward-and-back motion which will wash off the surface grains a few at a time. The washing can be effectively controlled by use of a somewhat circular motion as well as the forward-and-back dipping motion. When the panner decides that further washing would cause a loss of values, the bed is re-stratified and more light material brought to the surface by repeating the vigorous side-to-side motion. Repeat the washing and shaking operations until the heavy-mineral concentrate is clean or until it is reduced to a volume small enough to permit inspection or removal of the gold. During the finishing steps the panner can save time by raking any remaining pebbles out of the pan with his fingers and flicking out smaller particles with his thumb. These and other tricks come with practice.

e.   *Inspection and estimating:* At the end of the panning operation the original material will normally have been reduced to a small quantity of concentrate consisting mostly of black sand minerals. After putting a little clear water in the pan, the experienced panner will fan out the concentrate on the bottom of the pan and by "tailing" the gold he can inspect or count the colors. At this point he can estimate the tenor of the sample. There are perhaps as many ways of tailing the gold as there are panners but this is usually accomplished by moving the pan in a way that causes the water to gently swirl around the trough formed at the intersection of the bottom and side of the pan. This swirl of water carries the lighter particles ahead of those which are heavier or finer and with careful manipulation, brings the gold colors into view at the tail of the slowly moving fan of concentrate.

f.   *Removing the gold:* Final separation of the gold from other heavy minerals can be made in a number of ways. Larger pieces can be picked out with tweezers or the point of the knife and small colors or specks can be picked up by pressing down on them with the end of a wooden match or a dry finger tip. Remove the gold by placing the finger tip over a vial of water and washing it off with a splash of water.

A small globule of clean quicksilver (mercury), if rolled around in the pan, will pick up the gold providing it is untarnished and free of oil or grease. Tarnished gold can be brightened by rubbing it in the pan.

Where there is a considerable amount of black sand, particularly if it is fine and densely packed, it may be easier to separate the gold by blowing. This is done by first drying the concentrate and then removing the magnetite with a common horseshoe magnet, and finally, by blowing the non-magnetic black sand residue away from the gold. To do this, place the non-magnetic residue in a dry gold pan or on a suitable sheet of stiff paper and holding it level, blow gently across its surface while tapping the pan or paper. This takes some practice but with care a clean separation can be made in a surprisingly short time.

2. **GENERAL NOTES ON PANNING WITH SUGGESTIONS FOR IMPROVING TECHNIQUE**

   a. *New pan – preparation for use:* The film of grease or other rust preventative found on a new pan *must be removed* before use. This is best done by passing the pan over a gas stove burner or other suitable flame until the metal turns blue. Although this process is sometimes called "burning", care should be taken to avoid excessive heat. Blueing a pan not only frees it of grease but equally important, the resulting dark color makes fine specks of gold much easier to see in the pan. The "burning" process should be repeated as often as necessary to keep the pan free of body oil films which accumulate on a pan in normal use.

   b. *Pan factor:* Gold pans are made in a variety of sizes but the size generally referred to as "standard" has a diameter of 16 inches at the top, 10 inches at the bottom and a depth of 2½ inches. A typical pan will hold 336 cubic inches or 0.0072 cubic yard. The number of pans representing a cubic yard of material in place (bank-measure) is called the pan factor. Pan factors vary according to the size and shape of the pan, the swell of the ground when excavated, and the amount of heaping when filling the pan. In practice, factors for a 16-inch pan range from 150 to 200 but an approximate figure of 180 is often used. This is based on a struck pan (i.e., level full) and an assumed 20 to 25 percent gravel swell.

   c. *Recommended pan size:* The average panner should not use a 16-inch pan but instead should use the so-called "half-size" pan which has a top diameter of 12 inches, a bottom diameter of 7½ inches and a depth of 2 inches. The half-size pan level-full weighs approximately 9 pounds compared to about 20 pounds for the standard 16-inch pan. Halving the sample weight by use of the smaller pan not only reduces fatigue, particularly when much panning is to be done, but the fact that it is much easier to carry in the field and to use in a small stream or tub is an important consideration. The pan factor for a 12 x 7½ x 2-inch pan is about 400, assuming a 20 to 25 percent gravel swell. Experience has shown that two half-size pans can usually be washed in less time than one full-size pan.

   d. *Use of perforated pan:* Panning, at best, is a tedious, back-breaking job and anything done to speed the operation or improve working conditions will be repaid many times over in the form of more reliable results. The beginner and experienced panner alike can profit by use of a sieve[1] made by drilling ¼-inch holes in the bottom of a pan of the same size and shape as the one used for panning. To use the sieve, place it inside of the regular pan and then fill with gravel and submerge in water in the usual way. When the material is thoroughly wetted, lift the sieve slightly and twist it back and forth (under water) until all minus ¼-inch material has passed into the regular pan. The plus ¼-inch material is discarded and the fines which dropped into the regular pan are washed in the usual way. Aside from speeding the overall panning operation, the use of a sieve enables the engineer to conveniently inspect the plus ¼-inch rocks and to estimate the proportion of coarse material.

e. *Use of safety pan:* Allowing the pan tailings to fall into a second pan generally referred to as a "safety" pan will guard against losing the sample by accident and will greatly expedite repanning where this is called for.

f. *Panning large samples:* When a large multi-pan sample is to be washed, rather than complete each successive pan, it is best to reduce them only to a rough concentrate. The rough concentrates are accumulated and are eventually combined for finishing in the usual manner.

g. *Stage panning:* Where a large amount of heavy black sand is encountered, a stage-panning procedure can be used to advantage. This is done by panning and repanning to successive high-grade concentrates without attempting to make a complete saving of black sand or values at any one stage. As the proportion of heavy minerals decreases with each successive repan, it becomes progressively easier to make a high-grade concentrate with a low-grade tailing. Usually two or three repannings will make an acceptably clean tailing.

h. *Supplemental data:* When panning a sample the experienced engineer will note a variety of things among which are: Its amenability to washing, particularly where clay or cementing materials are present; the proportion of coarse to fine material; any evidence of unusual muddy water problems; the composition and angularity of rocks; the relative ease of concentration; the quantity and composition of black sand; indications of valuable or potentially valuable accessory minerals; the size, shape and other physical characteristics of the gold including "rust", tarnish or other factors which would affect its amalgamation. Any of the foregoing could be important factors in a placer mining operation.

i. *Use as a geologic tool:* Although the miners' pan is normally associated with gold deposits, it can be profitably employed when investigating a variety of heavy minerals such as monazite, scheelite, magnetite,

---

1/ Sometimes referred to as a "grizzly" pan.

90

ilmenite, cassiterite, chromite, cinnabar, etc. In general, it should be borne in mind that with few exceptions the pan can be employed in the study of either lode or detrital-type deposits containing finely divided minerals of relatively high specific gravity. The use of a miners' pan as a geologic tool has been studied and reported in detail by Mertie (1954) and by Theobald (1957).

3. FIRE ASSAY OF PLACER SAMPLES – MISLEADING RESULTS

Fire assaying, in essence, is a minature smelting process which recovers and reports the *total* gold content of the assay sample, including gold combined with other elements or locked in the ore particles. Because of this, a fire assay may report values that cannot be recovered by placer methods and it cannot be too strongly stressed that when dealing with gold placers, the sample values should *not* be determined by fire assay. Furthermore, no credence should be placed in placer valuations or reports that are based on the results of fire assays. Although this should be common knowledge among mineral examiners, a surprising number seem unaware that fire assaying although accurate per se yields misleading results when applied to placers.

There are other reasons: First, consider the small quantity of material used in a fire assay. The usual amount of sample taken for a crucible charge is either 29.166 grams (one assay-ton) or half of this amount. Next, consider that a particle of placer gold only 1/32-inch in diameter may weigh about 1/4 milligram. Now if the bank-run material from which it came averaged 10c per cubic yard, [1] the 1/4-milligram gold particle would be contained in about 7½ pounds of sand and gravel. But suppose this same 1/4-milligram gold particle found its way into a 29-gram fire assay charge. The resulting assay-value would be 1/4 ounce per ton or about $5.80 per cubic yard.[1] It is seen that a single small particle of gold, by placer standards, will cause an intolerable error when injected into a standard fire assay charge.

Experience tells us that with few exceptions no amount of mixing or careful division can produce a fire assay charge representative of bank-run placer material. The practice of first panning a sample to reduce its bulk, and then fire assaying the resultant black sand concentrate does not entirely resolve this problem.

Even if we assume that a representative crucible charge could be obtained, the fire assay will detect *all* gold including that which is locked up in rock particles or is too finely divided to be recovered by placer methods. This in itself precludes the use of fire assays for evaluating placer ground.

In brief, experience has shown that fire assay results applied to placers usually results in a substantial over-valuation of the ground. This fact has been set out by Janin (1918, p.38), Vanderburg (1936, p. 33), Gardner and Allsman (1938, p.61) and others.

1/ Based on gold at $35 per ounce.

4. PROCEDURE FOR DETERMINING RECOVERABLE GOLD IN PLACER SAMPLES

a. Reduce the original sample to a black sand concentrate by panning, rocking, or other suitable means.

b. Place the concentrate in a pan, then count and record the gold colors as #3, #2 or #1 – size (Note 1). At this point, manually remove any gold particles which are to be kept in their natural form (Note 2).

c. Add a globule of clean, gold-free mercury (about the size of a small bean) and working over a safety pan reduce the black sand to a small-volume concentrate. Near the end of panning process the mercury may tend to ride over the top of fine-size or hard-packing materials but by this time all gold should have been picked up by the mercury.

d. Remove the mercury and place it in a 250 ml Pyrex beaker. Add 40 or 50 ml diluted nitric acid (Note 3) and digest until the mercury globule is reduced to the size of a match head. Transfer to a #0 glazed porcelain parting cup, add fresh acid and complete the digestion using low heat if necessary. Fine-size gold will be left as a coherent, sponge-like mass if rapid digestion or boiling is avoided.

e. After decanting off the acid, carefully wash the gold three or more times with warm water. Add a drop or two of alcohol (to prevent spattering) and dry in the parting cup at low heat.

f. Anneal the gold by bringing the bottom of the parting cup to a low red heat. This step will eliminate any residual mercury and is essential when working with very small gold weights.

g. Transfer the annealed gold to a balance pan and weigh. A balance widely used by placer engineers is shown in Figure 20.

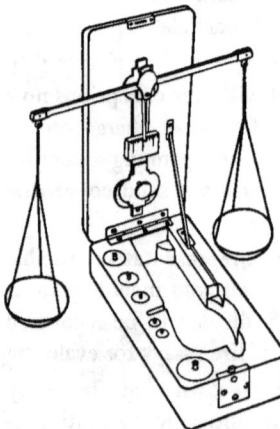

FIGURE 20 – A folding assay balance widely used in placer work. Capacity 10 grams, sensitive to ¼ milligram. Complete in mahogany case with set of weights 10 grams by 1 milligram. Size when closed, about 6x3x1½ inches.

NOTES

1. Number 3 colors consist of gold particles weighing less than 1 milligram. Number 2 colors weigh between 1 milligram and 4 milligrams. Number 1 colors weigh over 4 milligrams. Gold particles weighing 10 milligrams or more should be individually weighed and recorded.

2. Placer gold loses its original surface texture when amalgamated. In the case of fine-size gold, its end form may assume the shape of fine needle-like crystals which are actually pseudomorphs after gold amalgam.

3. Prepare a parting solution by diluting commercial nitric acid with water to a strength which will dissolve mercury without violent action. A 1:1 dilution is usually satisfactory.

4. Large-volume or hard-packing black sands that are difficult or tedious to pan may be amalgamated in a glass jar rotated about its longitudinal axis at a peripheral speed of 500 inches per minute. The addition of a small amount of lye will counteract oil or grease and will generally assist amalgamation.

5. The described procedure will extract all free gold recoverable by commerical placer methods. With care, extremely low gold values can be determined. Where so-called "rusty" or coated gold is present, provisions should be made to scour or otherwise brighten the gold prior to amalgamation. This may be done by rubbing it on the bottom of the pan or adding a few pebbles to the amalgamating jar. Fine grinding or pulverizing should be avoided.

REFERENCES CITED (PART V)

Gardner, E. D. and Allsman, Paul T., 1938, Power-shovel and dragline placer mining: U.S. Bureau of Mines Information Circular 7013, 1938.

Janin, Charles, 1918, Gold dredging in the United States: U.S. Bureau of Mines Bulletin 127, 1918.

Mertie, John B. Jr., 1954, The gold pan; a neglected geologic tool: Econ. Geology, Vol. 49, No. 6, Sept. – Oct., 1954, pp. 639-651.

Theobald, Paul K. Jr., 1957, The gold pan as a quantitative geologic tool: U.S. Geol. Survey Bull. 1071-A, 1957.

Vanderburg, William O., 1936, Placer mining in Nevada: University of Nevada Bull. Vol 30, No. 4, May 15, 1936.

PART VI

NOTES ON GENERAL PRACTICE

1. DESCRIPTION OF SAMPLES

A mining claim may be validly located and held only after the discovery of a valuable mineral deposit. When it becomes necessary to determine the validity of a claim and the sufficiency of discovery is in doubt, the Government will initiate contest proceedings during which the testimony is given by parties to the contest and their witnesses. During such proceedings, assay certificates or other sample data are submitted as evidence of value, or lack of value. The sample and assay data are important in that the hearings examiner or the court must rely on them to a large extent to determine if a valid discovery has been made.

But numerous decisions have pointed out that in order to fully understand the results of these assays, the method in which samples were taken and tested must be considered. Without such information they have little probative value and, accordingly, are entitled to little evidentiary weight.

In brief, an accurate and systematic record should be kept of each sample through the cutting, processing, and calculation stages. The use of suitable printed forms will expedite the data recording process, and, equally important, will serve as a check list and lessen the risk of oversight. Three such forms prepared specifically for BLM use are included in the appendix.

Complete, accurate and signed sample records are in effect legal documents and they should be prepared and preserved with corresponding care.

2. REDUCTION OF SAMPLE VOLUMES

In typical gold placers the variations are so great and the values are so low, any attempt to divide a sample by taking alternate shovels, mechanical splitting, or by other means will invariably yield erratic results. Rather than set out supporting theory which at best would be academic, two examples are offered to show what actually happens. One is based on a laboratory experiment and the other on field practice at an operating mine. First consider the laborabory experiment.

Referring to Figure 21, assume that the bucket of sand weighs 60 pounds and that instead of lead shot it contains 320 particles of placer gold weighing 0.1 milligram each.

The initial sample will have a value of $1.60 per cubic yard.[1]

The first split yields two 30-pound samples with indicated values of $1.53 and $1.67 per cubic yard, a spread of 14 cents per cubic yard.

---

[1] Using a weight to volume ratio of 3,000 lbs. per cu. yd. and figuring 10 milligrams of gold to the cent.

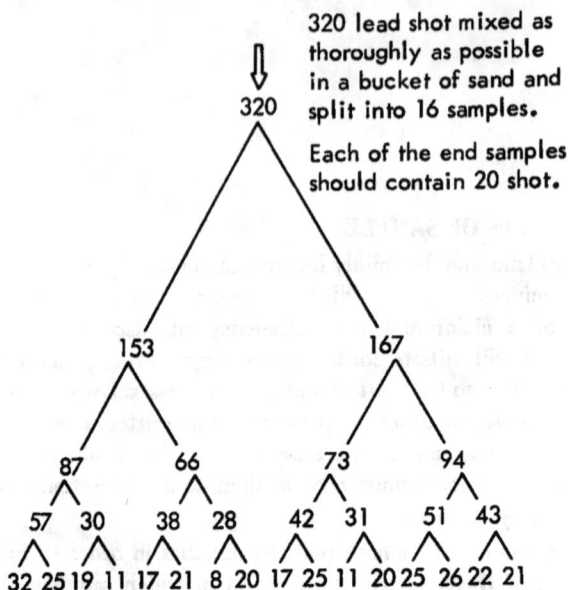

320 lead shot mixed as
thoroughly as possible
in a bucket of sand and
split into 16 samples.

Each of the end samples
should contain 20 shot.

*FIGURE 21 — Diagram showing erratic distribtuion of metallic
particles in successive splits of a placer-type sample.*

The second split yields four 15-pound samples with indicated values of
$1.74, $1.32, $1.46 and $1.88 per cubic yard; a spread of 56 cents per cubic
yard.

The third split yields eight 7½-pound samples with values ranging from a low
of $1.12 to a high of $2.28 per cubic yard; a spread of $1.16

The fourth split yields sixteen 3-3/4-pound samples with values ranging from
a low of 64 cents to a high of $2.56 per cubic yard; a spread of $1.92.

As splitting continues, the spread in indicated value becomes progressively
greater. On the fifth split, for example, (not shown) the indicated values
ranged from 16 cents to $2.72 per cubic yard.

Data from a recent study by the United States Geological Survey (Clifton,
1967) confirm that variances shown in the foregoing example are typical of
those found in practice.

The second example has been taken from the prospecting records of an
operating company. In this case the gold-bearing gravel excavated from each
of 14 shafts was divided as carefully as possible by taking alternate buckets
as the work progressed. The hand-dug shafts were 40 inches in diameter and
ranged from 30 to 40 feet deep. The individual halves, averaging 7¼ cubic

yards each, were washed separately. Results are shown in the following tabulation:

Shaft No.	Value c per cu. yd.	Indicated Value		% of Actual Value	
		Half "A"	Half "B"	Half "A"	Half "B"
1	27.0	17.4	36.6	64	136
2	26.7	24.8	28.7	92	108
3	31.0	35.0	27.0	113	87
4	34.6	33.7	35.6	98	102
5	31.0	26.6	35.5	86	114
6	13.0	13.2	12.8	102	98
7	38.3	36.5	40.2	95	105
8	54.9	61.2	48.6	112	88
9	19.1	17.7	20.6	92	108
10	26.5	24.0	29.0	90	110
11	25.3	23.8	26.8	94	106
12	29.5	26.6	32.4	90	110
13	26.8	28.0	25.6	104	96
14	15.7	15.2	16.2	97	103

TABLE III. — *Erratic values resulting from 1:1 splits of gold-bearing gravel. Two equal-volume samples were obtained from each shaft by placing alternate buckets in half "A" and half "B".*

The foregoing examples show that the splitting of gold placer samples should be strictly avoided unless there is some very special reason to divide a sample or reduce its volume before washing or other treatment. In cases where this *must* be done, the mineral examiner should be fully aware of the erratic and probably misleading results likely to follow. Any divergence from normal procedure should, of course, be noted in the sampling records and in subsequent reporting.

When dealing with fine-size placer materials having a low unit-value, such as magnetite or ilmenite sands, a substantial reduction of the bulk sample or of its concentrate by mixing and splitting is generally a proper procedure and a common practice.

3. REPORTING VALUES

In the United States, the volume of placer material is always reported as bank-measure cubic yards and the value is customarily reported as cents per cubic yard.

In addition to the calculated value, the actual weight of gold in milligrams, its fineness or estimated fineness, and the price of gold used in the value calculation should be noted on the sample sheets as well as in subsequent reporting.

Minerals other than the precious metals are reported as pounds per cubic yard, percent, or the particular unit customarily used for the commodity in question.

4. COST ESTIMATES

One of the mineral examiner's functions is to give a qualified or expert opinion as to whether a given mining claim or mineral deposit has present or potential value. Among the evidence to be examined and weighed in forming his judgement will the question of costs. In many cases there is a factor so obviously weak that a detailed cost study is not necessary, in fact, it would be useless. Here, a rough approximation based on experience and common sense will usually suffice.

But in other cases a careful consideration of this question will be necessary. If the deposit has not been completely blocked out and sampled, or if the mining method and working rate have not been firmly established, an accurate cost estimate is not possible. In the absence of such development work, all the mineral examiner can do is to make a qualified estimate based on comparative data from reliable sources, or from his own experience with similar properties.

Basically, he must decide whether the deposit contains enough gold or other salable mineral to cover:

> Cost of operation.
>
> Cost of all equipment.
>
> Royalties or cost of land.
>
> All exploration, sampling, and pre-operating costs.
>
> Cost of borrowed captial.
>
> Cost of start-up delays and contingencies.
>
> Reasonable profit.

In turn, some of the factors influencing the operating cost will be:

> Type of operation (dredge, hydraulic, sluiceplate, etc.)
>
> Size and tenor of the deposit.
>
> Character of the gravel and bedrock.
>
> Water supply.
>
> Grade for sluices.
>
> Disposal area for tailings and muddy water.
>
> Working season.
>
> Regulatory laws (land restoration, etc.)

If the foregoing are kept in mind and weighed intelligently, a fair cost estimate can usually be made by comparing the subject property with a similar property where mining operations have been conducted and costs established. But just how to obtain these cost data and how to equate them to today's inflated prices is one of the practical problems encountered in applying the comparative cost method.

First, we find that because of the depressed state of the placer mining industry there are few operations from which up-to-date costs might be extracted. Second, anyone who has tried to obtain detailed cost data from a responsible operating company has soon found that such information is considered confidential and seldom released for outside use. While the small operator may be freer with his information, it will usually be found that he did not bother to keep systematic records and his cost figures may be little more than a guess.

This means that for the most part, the mineral examiner will have to rely on the few placer cost data available in the technical press, most of which are fragmental and over 30 years old. A list of the more useful and readily available sources has been placed at the end of this section.

To update the older figures, a year-by-year cost index will be needed. The annual index for various classes of construction, construction materials, labor, and machinery can be obtained from Engineering News-Record magazine. The March (quarterly review) issue will be found particularly helpful.

In the absence of comparative data the estimator is on his own. In such cases he may have to work up estimated costs based on a hypothetical operation using a method or combination of methods proven effective in similar deposits elsewhere. This estimating procedure has been illustrated by Staley and Storch in a two-part article titled "Choosing a mining method for gold-bearing gravels." This article has been listed among the selected references which follow.

Examples of 1964 costs are given in Appendix E.

SELECTED REFERENCES CONTAINING PLACER COST DATA

Averill, Charles V., Dragline dredging in Siskiyou County: Thirty-seventh Report of (Calif.) State Mineralogist, pp. 328-331. April, 1949. Costs of principal items of equipment noted. Also cost of a resoiling project.

Bureau of Mines Staff, Minerals Yearbook, annual volumes. Current economic trends and effects on mining industry.

— Production potential of known gold deposits in the United States: U.S. Bureau of Mines Information Circular 8331, pp. 11-14, 1967. Notes current cost estimates for dredging, drift-mining and hydraulicking. Also capital cost for bucket-line dredges.

Chellson, Harry, C., What will it cost to work a gold placer of medium size? Engineering and Mining Journal, Vol. 135, No. 10, pp. 441-445. October, 1934. Costs for various size operations using a mobile dryland washing plant.

Daily, A.F., Chapter 13.5 (Placer Mining) in AIME Surface Mining volume: Published by Amer. Inst. of Mining, Metallurgical and Petroleum Engineers, Inc., New York. 1968.

Eassie, W. H., What does hydraulicking cost? Engineering and Mining Journal, Vol. 145, No. 6, pp. 92-94. June 1944. 1939-1941 costs for a medium-size hydraulic mine in British Columbia, Canada.

Gardner, E. D., and Allsman, Paul T., Power-shovel and dragline placer mining: U.S. Bureau of Mines Information Circular 7013, 68 pp., 1938. Detailed cost data for floating and stationary plants.

Gardner, E. D., and Johnson, C. H., Placer Mining in the Western United States. I. General Information, hand-shoveling, and ground sluicing: U.S. Bureau of Mines Information Circular 6786, 73 pp. 1934.

___ Placer mining in the Western United States. II. Hydraulicking, treatment of placer concentrates, and marketing of gold: U.S. Bureau of Mines Information Circular 6787, 88 pp. 1934. Contains operating and cost data on small-scale hydraulicking.

___ Placer mining in the Western United States. III. Dredging and other forms of mechanical handling of gravel, and drift mining. U.S. Bureau of Mines Information Circular 6788, 81 pp., 1935.

Lord, Harry S., Modern dragline dredging: Mining World, Vol. 3, No. 10, pp. 14-16. Oct., 1941. Cost of principal items of equipment for 1½ cu. yd. to 12 cu. yd. dragline dredges.

Magee, James F., A successful dragline dredge: A.I.M.E. Trans., Vol. 126 (1937), pp. 180-200. Contains detailed operating data for a typical dragline dredge.

Patman, Charles G., Methods and costs of dredging auriferous gravels at Lancha Plana, Amador County, California: U.S. Bureau of Mines Information Circular 6659, 17 pp. 1932. Describes construction, operation, and operating costs of a 6-cu. ft. bucket-line dredge.

Romanowitz, Charles M., Floating dredges used for mining purposes: Mineral Information Service, Vol. 20, No. 7, pp. 82-87. July, 1967. (Published by California Division of Mines and Geology). Discusses onshore and offshore dredging including tin and other minerals. Compares hydraulic (suction) dredges and bucket-line dredges. Notes dredging costs in South America and Yukon Territory. Reviews history of dredging and latest developments.

Staley, W. W., and Storch, R. H., Choosing a mining method for gold-bearing gravels, Part I: Engineering and Mining Journal, Vol. 138, No. 7, pp. 342-346; also p. 359, July, 1937. Part II: Vol. 138, No. 9, pp. 43, 44. Sept., 1937. Develops hypothetical mining programs and comparative costs for working a placer (in Alaska) by hydraulicking, power drag-scraper, and dredging. A good example of a placer cost study.

Thomas, Bruce I., Cook, Donald J., Wolff, Ernest, and Kerns, William H., Placer mining in Alaska: U.S. Bureau of Mines Inforamtion Circular 7926, 34 pp. 1959. Describes mining methods and gives unit operating costs at operations where hydraulic and mechanical equipment is used with nonfloating washing plants, including sluiceplate mining operations.

Thurman, Chas. H., Costs in dragline gold dredging: A.I.M.E. Technical Publication No. 1900 (1945), 6 pp. Compares bucket-line and dragline dredge costs. Gives examples of equipment and operating costs.

5. UNPROVEN PROCESSES

Many special placer machines or secret recovery processes have been "invented" or proposed. Some claim the ability to extract microscopic or colloidal gold from materials that show little or no value when tested by fire assay or by the normal methods of testing placer material. Others are intended to recover the varying amounts of fine gold admittedly lost in large-scale placer operations. These devices or schemes seem to have an unfailing attraction for miners as well as for the general public.

But despite the many "improved" placer machines and the new gold-saving methods that have been offered, the simple Hungarian-type riffle has held its place in the placer industry while most of its rivals have been discarded. A notable exception is the placer-type jig which has supplemented the riffled tables (sluices) in some dredging operations and has replaced them in others.

It can be said generally that the success of a placer operation will hinge on the throughput, a high throughput being the key to low costs. In other words, the greater the throughput the lower the unit cost. Experience has shown that to achieve the optimum working rate some recovery must be sacrificed, or put simply, it is cheaper to lose a certain amount of gold than to save it.

To date no one had demonstrated a placer recovery system that can economically replace today's methods and equipment. Operating economies made possible by the large capacity and the simplicity of conventional riffles and placer jigs more than offset the dollar value of the gold they may lose. On this basis, they yield the greatest operating profit.

Even where a new or improved recovery method may be shown to possess some potential, if it is not yet at the stage where it can be presented as a proven method or technique, the mineral examiner has little choice but to rely on standard anaytical and recovery methods when making his evaluation.

REFERENCES CITED (PART VI)

Clifton, H. Edward, Hubert, Arthur, and Phillips, R. Lawrence, Marine Sediment Sample Preparation for Analysis For Low Concentrations of Fine Detrital Gold: U.S. Geol. Survey Circular 545, 1967. 11 pp.

PART VII

CHECK LIST FOR PLACER INVESTIGATIONS

1. GENERAL CONSIDERATIONS

An adequate mineral investigation will develop a considerable body of information in addition to that obtainable from samples alone, some of which will be found vital when assessing the actual worth of a prospect. For example: Sometimes most of the value in a commercial placer is found on or in the bedrock; perhaps several feet of bedrock must be taken up to recover the pay. In such cases its hardness and irregularities must be known and failure to consider this has proved fatal to more than one dredging project. Because boulders can be disastrous to a placer operation their maximum size, number, and distribution should always be of prime concern to the mineral examiner. Failure to recognize or properly assess tailings or muddy water disposal problems, where they exist, can prove expensive or cause premature shutdown of an otherwise profitable operation. In brief, physical details often determine the success or failure of a placer as much or more than the mineral content itself. The importance of considering *all* factors and their possible effect cannot be overemphasized.

Because no two deposits are alike, no rule can be made as to what should be included in a placer check list. Also, the degree of inquiry will vary widely depending on the purpose of the investigation. Where it is clearly evident that a property has no value or prospective value a detailed field investigation may not be required, but even here, sufficient data for a well-informed report should be gathered.

The following check list is intended first as a field guide and second, to show the range of inquiry which may be necessary for an adequate placer investigation. The user should, of course, tailor this list to suit his particular needs.

2. FIELD GUIDE AND CHECK LIST FOR PLACER INVESTIGATIONS

1. Date of examination _____

2. NAME OF CLAIM(s) OR PROPERTY _____

_____

_____

3. State _____ , County _____ District _____

4. Township _____ , Range _____ , Section(s) _____

5. REASON FOR EXAMINATION _____

_____

_____

6. EXAMINED BY _____

7. Assisted by _____

8. Others present _____

_____

_____

9. Number of claims or acres _____

_____

10. NAMES OF LOCATORS AND PRESENT OWNER _____

_____

_____

11. Owner's Address _____

_____

12. TYPE OF DEPOSIT (stream, bench, desert, etc.) _____

_____

_____

13. Terrain _____

_____

14. Gradient of deposit: Less than 5% ( ); More than 5% ( ).
    Remarks _____

_____

15. Is the deposit dissected by deep washes or old workings? Yes ( ); No ( ).

Remarks _____

_____

16. Type and extent of overburden _____

_____

_____

17. Depth to permanent water table _____

_____

18. Depth to bedrock _____

_____

_____

19. Kind of bedrock (rock type) _____

_____

20. Hardness of bedrock _____

_____

21. Bedrock slope or contour to be expected _____

_____

_____

_____

22. Are high bedrock pinnacles or reefs in evidence? Yes ( ); No ( ).

Remarks _____

_____

_____

23. Gravel is: Well-rounded ( ); Sub-rounded or Sub-angular ( ); Angular ( ).

Remarks _____

_____

24. Does gravel contain rocks over 10-inch ring size? Yes ( ); No ( ).

Remarks _____

_____

25. BOULDERS (Max. size, number, distribution, etc.) _____

_____

_____

_____

26. Rock types noted in gravel _____

_____

_____

_____

27. Predominent rock type (if any) _____

_____

28. SAND (kind, amount, distribution, etc.) _____

_____

_____

_____

29. Sorting or bedding patterns (if apparent) _____

_____

_____

_____

30. STICKY CLAY? Yes ( ); No ( ). Remarks _____

_____

31. Cemented gravel? Yes ( ); No ( ). Remarks _____

_____

_____

32. Caliche? Yes ( ); No ( ). Remarks _____

_____

_____

33. Permafrost? Yes ( ); No ( ). Remarks _____

_____

34. Buried timber? Yes ( ); No ( ). Remarks _____

_____

110

35. Hard or abrasive digging conditions? Yes ( ); No ( ). Remarks _____

_____

_____

_____

36. Character of gold: Coarse ( ); Flaky ( ); Fine ( ); Rough ( ); Shotty ( ); Smooth ( ); Bright ( );

Stained or coated ( ). Remarks _____

_____

_____

37. Can good recovery be expected by use of riffles or jigs? Yes ( ); No ( ). Remarks _____

_____

_____

38. Is recovery said to depend on secret process or special equipment? Yes ( ); No ( ). Remarks _____

_____

_____

_____

39. Are black sands said to contain locked gold values? Yes ( ); No ( ). Remarks _____

_____

40. Have black sands been checked for valuable minerals other than gold? Yes ( ); No ( ).

Remarks _____

_____

_____

41. Distribution of values in deposit (if known) _____

_____

_____

_____

42. Record or evidence of previous sampling _____

_____

_____

_____

43. Results of prior sampling (if known) _____

_____

_____

44. Are old workings in evidence? Yes ( ); No ( ). Remarks _____

_____

_____

45. Past production (if known) _____

_____

_____

_____

46. Date of last production or work _____

_____

47. Reason for quitting _____

_____

48. Present work (if any) _____

_____

_____

_____

_____

49. APPLICABLE MINING METHOD _____

_____

_____

_____

50. Possible cost to bring property into production _____

_____

_____

_____

_____

51. POSSIBLE MINING COST _____

_____

_____

52. Dimensions of (physically) minable ground _____

_____

_____

53. Possible extensions _____

_____

_____

_____

54. Maximum yardage indicated to date _____

_____

_____

55. Mining equipment on ground _____

_____

_____

56. Accessory equipment or improvements on ground _____

_____

_____

_____

_____

57. Water supply _____

_____

_____

_____

_____

58. Power supply _____

_____

_____

_____

59. DOES PROPERTY HAVE ADEQUATE TAILINGS DUMP ROOM? Yes ( ); No ( ). Remarks _____

_____

_____

_____

113

60. Would mining in this area come under County, State or Federal water Quality control regulations?

Yes ( ); No ( ). Remarks _____

_____

_____

_____

61. Fish and Game regulations? Yes ( ); No ( ). Remarks _____

_____

_____

62. CAN SETTLING PONDS BE BUILT TO EFFECTIVELY RETAIN OR CLARIFY THE MUDDY WATER? Yes ( ); No ( ). Remarks _____

_____

_____

63. IS PROPERTY SUBJECT TO RESOILING OR OTHER SURFACE RESTORATION REGULATIONS?

Yes ( ); No ( ). Remarks _____

_____

_____

_____

64. Elevation of property _____

65. Climate _____

_____

66. Working season _____

67. Season governed by _____

_____

68. Surface cover and its effect on mining _____

_____

_____

69. Merchantable timber or other surface values _____

_____

_____

_____

70. Nearest town _____

114

71. Access _____

_____

_____

72. Reference maps _____

_____

_____

73. Aerial photos (USGS, Forest Service, etc.) _____

_____

_____

74. Reference literature _____

_____

_____

75. Previous examinations or reports _____

_____

_____

76. Other reference sources _____

_____

_____

77. SAMPLING (describe or attach notes) _____

_____

_____

_____

78. Additional information and remarks _____

_____

_____

_____

79. Attach suitable map or sketches (if needed). _____

80. Attach photographs of pertinent features (if available).

115

PART VIII

GLOSSARY OF
PLACER TERMS

Many of the following terms have universal definitions, that is, they have definitions common to all branches of the mineral industry. On the other hand some are unique to the placer industry or at least they have placer-related meanings different from those in general use. For example: The term FLOTATION, as used in the general mining industry, relates to a mineral separation process. But in placer mining, the term FLOTATION is applied to the minimum water depth needed to move an operating dredge.

It should also be noted that the definitions given here are intended to be descriptive rather than legal, and they should be used accordingly.

Names in parentheses refer to sources as follows:

AGI. Dictionary of Geological Terms. Prepared under the direction of the American Geological Institute. Doubleday & Company, Garden City, N.Y. 1962.

Brooks. Brooks, A. H., The Gold Placers of Parts of Seward Peninsula, Alaska. U.S. Geol. Survey Bull. 328, pp. 114-145, 1908.

Dunn, E. J. Dunn, E. J., Geology of Gold. Charles Griffin & Co., London. 1929.

Dunn, R. L. Dunn, Russell L., Drift Mining in California. In Eighth Report of the (Calif.) State Mineralogist, pp. 736-770. 1888.

Fay. Fay, Albert H., A Glossary of the Mining and Mineral Industry. U.S. Bureau of Mines Bull. 95. 1920.

McKinstry. McKinstry, Hugh E., Mining Geology. Prentice-Hall, New York. 1948.

ACCRETION BAR. A low-level deposit of sand and gravel formed in a stream by gradual addition of new material. Accretion bars are typically formed along the short, or inside radius of curves. See - Skim bar.

ADJUSTED VALUE. A sample value that has been increased or decreased by an amount deemed necessary to offset known variables or other factors that may cause discrepancies in the initially indicated value. In placer drilling, the adjusted value is also known as a CORRECTED VALUE. To be valid, such adjustments must be based on careful diagnosis of sampling problems, and must reflect sound judgement. See — Indicated value.

AINLAY BOWL. A wet, gravity concentrator used for the recovery of gold and other heavy minerals from alluvial materials. It consists essentially of a bowl-shaped vessel, rotated about its vertical axis and provided with circular riffles. Feed entering at the center is carried upward and outward by the flow of water and centrifugal force. Tailings overflow the rim while gold and other heavy minerals are retained by the riffles. A somewhat similar bowl-shaped concentrator is known as the KNUDSEN BOWL.

AIRPLANE DRILL. A compact, engine-powered placer drill designed for use in areas of difficult access. The term AIRPLANE DRILL is actually a trade name which through common use, has become part of the placer vernacular.

ALLUVIAL. 1. Deposited by a stream. 2. Relating to deposits made by flowing water. (Fay)

ALLUVIAL FAN. A cone-shaped deposit of alluvium made by a stream where it runs out onto a level plain or meets a slower stream. The fans generally form where streams issue from mountains upon the lowland. (AGI)

ALLUVIAL GOLD. Gold found in assocation with water-worn material.(Fay)

ALLUVIAL PLAIN. 1. Flood plains produced by the filling of a valley bottom are alluvial plains and consist of fine mud, sand, or gravel. 2. A plain resulting from the deposition of alluvium by water. (AGI)

ALLUVIUM. A general term for all detrital deposits resulting from the operations of modern rivers, thus including the sediments laid down in river beds, flood plains, lakes, fans at the foot of mountain slopes, and estuaries. (AGI)

AMALGAM. An alloy of mercury with gold or another metal. In the case of placer gold, a "dry" amalgam, that is, one from which all excess mercury has been removed by squeezing through chamois leather will contain nearly equal proportions of gold and mercury.

AMALGAMATION. The extraction of the precious metals from their ores by treatment with mercury.

ANCIENT BEACH PLACER. Deposits found on the coastal plain along a line of elevated beaches. (Brooks)

ANCIENT CHANNEL. See Tertiary channel.

ANNUAL LABOR. See Assessment work.

ASSAY.(verb) To determine the amount of metal contained in an ore. (McKinstry)
    1. The act of making such a determination.
    2. The result of such a determination.
    See – Fire assay.

ASSAY VALUE. The amount of gold or silver, contained in an ore or other material, as shown by assay of any given sample.

ASSESSMENT WORK. The annual work upon an unpatented mining claim on the public domain necessary under the United States law for the maintenance of the possessory title thereto. Same as ANNUAL LABOR. (Fay)

AURIFEROUS. Containing gold.

BAJADA PLACER. Placers found in confluent alluvial fans along the base of a mountain range or in a mantle of rock debris along the lower slope of a mountain range, in arid regions. The deposits are mainly residual detritus and poorly sorted alluvium found in gulches and on slopes that are subject to occasional torrential rain wash. Bajada is the Spanish term for slope. This term has not found general use in placer mining, most bajada placers being referred to collectively as "Desert" placers.

BANK-MEASURE. The measurement of material in place, such as gravel in a deposit before excavation. In placer work, values are normally reported as cents per cubic yard and unless specified otherwise, this means a cubic yard in place, or bank-measure.

BANK WATER. See By-Wash.

BANKA DRILL. A placer drill consisting essentially of a flush-jointed casing equipped with a serrated cutting shoe. The casing is rotated by means of a man or animal-powered sweep attached to the upper section. Men standing on an attached platform, chop up the drill core and remove it from the casing by means of hand-powered tools. Also known as an EMPIRE DRILL.

BAR. A deposit of alluvial material above or below the water line of present streams. Bars may form where the current slackens or changes direction. See — Accretion bar.

BATEA. A wide and shallow, cone-shaped vessel, usually of wood, used for panning gold. The batea is in common use in Mexico, Central and South America, and Asia.

BEACH PLACER. See Sea-beach placers.

BED LOAD. Soil, rock particles, or other debris rolled along the bottom of a stream by the moving water, as contrasted with the "silt load" carried in suspension. (AGI)

BEDROCK. The solid rock underlying auriferous gravel, sand, clay, etc., and upon which the alluvial gold rests. (Fay) In placer use, the term bedrock may be generally applied to any consolidated formation underlying the gold-bearing gravel. Bedrock may be composed of igneous, metamorphic or sedimentary rock. See — False bedrock.

BENCH PLACER. Gravel deposits in ancient stream channels and flood plains which stand from 50 to several hundred feet above the present streams. (Brooks)

BLACK GOLD. Alluvial gold coated by black oxide of manganese. (Dunn, E. J.)

BLACK SAND. Heavy grains of various minerals which have a dark color, and are usually found accompanying gold in alluvial deposits. (Fay) The heavy minerals may consist largely of magnetite, ilmenite and hematite associated with other minerals such as garnet, rutile, zircon, chromite, amphiboles, and pyroxenes. In Western gold placers, the black sand content is commonly between 5 and 20 pounds per cubic yard of bank-run gravel.

BLUE GRAVEL. Some of the deeper, water-saturated gravels found in California's Tertiary channels have a distinctive bluish-gray color and for this reason early miners referred to them as "blue gravel" or more commonly, as the "blue lead". At one time they were believed to represent a separate gravel flow, distinct from the overlaying red gravels. Actually, these blue gravels represent unoxidized portions of the gravel channels whereas the red gravels represent the oxidized portions of the same material.

BLUE LEAD. (pronounced leed) See — Blue gravel.

BOOMING. A variation of ground sluicing in which water is stored in a reservoir and suddenly released to provide a rush of water, in a large volume, which erodes and transports the gravel. Booming is generally employed where water is scarce. In California the contrivances for collecting and discharging water are termed SELF-SHOOTERS. See — Ground sluicing.

BRAIDED STREAMS. 1. A braided stream is one flowing in several divided and reuniting channels resembling the strands of a braid, the cause of division being the obstruction by sediment deposited by the stream. 2. Where more sediment is being brought into any part of a stream than it can remove, the building of bars becomes excessive, and the stream develops an intricate network of interlacing channels, and is said to be braided. (AGI) 3. Conditions which cause braiding are common in glacial areas where much sediment is added by the melting ice and in semiarid regions where the transporting power of streams is reduced by seepage and evaporation. In general, such conditions are not conducive to the formation of placers.

BREAKOUT. A point where a ravine or canyon cuts into, but not through, a channel. (Dunn, R. L.) Usually applied to buried Tertiary channels. Compare with Outlet, and with Inlet.

BREAST. The working face of a prospect drift on the pay lead; the face of a gangway being mined. (Dunn, R.L.)

BUCKET-ELEVATOR DREDGE. See — Bucket-Line dredge.

BUCKET-LINE DREDGE. A dredge in which the material excavated is lifted by an endless chain of buckets. (Fay) Also known as Connected-bucket dredge. The type of bucket-line dredge generally employed in placer mining is a self-contained digging, washing and disposal unit, operating in a pond and capable of digging, in some cases, more than 100 feet below water. Its machinery is mounted on a shallow-draft hull and the dredge backfills its working pit (pond) as it advances. The capacity of individual buckets is used as a measure of dredge size. For example, an "18-foot" dredge is equipped with buckets having a struck capacity of 18 cubic feet each. Compare with Dragline dredge; also Suction dredge.

BULLION. Unrefined gold that has been melted and cast into a bar. In placer mining, the gold sponge obtained by retorting amalgam, is commonly melted with borax or other fluxes, then poured into a bullion bar. See — Sponge.

BURIED PLACER. Old placer deposits which have become buried beneath lava flows or other strata. (Fay) See — Tertiary Channel.

BY-WASH. In many cases, hydraulic giants are capable of cutting more material from the bank, than can be swept into the slucies by means of the giants alone. In such cases supplemental water may be brought into the pit by means of a ditch, to assist carrying the material to the sluices. This is locally called BY-WASH, BY-WATER or BANK WATER.

BY-WATER. See — By-Wash.

CABLE DRILL. See — Churn drill.

CABLEWAY SCRAPER. See — Slackline scraper.

CAISSON. A metal cylinder used to sink prospect shafts in loose ground or in the presence of a large quantity of water. Caissons are usually provided in sets of 4 or more telescoping units.

CALICHE. A brown or white material commonly found as a subsoil deposit in arid or semi-arid climates, and which is composed largely of calcuium carbonate. It is commonly encountered in desert placers where its cementing effect adversely affects the mining and washing processes.

CANNON CONCENTRATOR. See — pinched sluice.

CAPPING. Volcanic flow materials or agglomerates that cover and in some cases, conceal underlying auriferous gravels. Commonly found associated with Tertiary channels in California's Sierra Nevada region. Also called CAP ROCK.

CASING. Steel tubing or pipe used to case a drill hole. In placer sampling it is usually driven into the formation ahead of the drill bit and when so used, is commonly called a "drive pipe".

CASING FACTOR. The depth to which a churn drill casing must be driven to take in a sample volume of 1 cubic yard. For example, a standard 6-inch drive pipe equipped with a new, 7½-inch drive shoe would be driven 88 feet to cut out a theoretical volume of 1 cubic yard. This is sometimes called PIPE FACTOR, but it is most commonly known as the DRIVE SHOE FACTOR. See — Radford factor.

CEMENT. The Material that binds together the sand and gravel particles in an indurated placer or other formation. The cementing material can be calcareous, silicious or ferruginous. Also used when referring to the hardened formations as a whole. Cemented gravels must, in some cases, be milled to release their gold content.

CEMENT CHANNEL. A channel depression completely filled with lava, no auriferous gravel. (Dunn, R. L.)

CHALK. Volcanic tuff or ash, largely rhyolitic in composition, is commonly found as intraformational strata or masses in Tertiary channels of California's Sierra Nevada region. The whiter, fine-grained and homogeneous beds are locally called "Chalk".

CHANNEL. A stream-eroded depression in the bedrock, ordinarily filled with gravel. See – Tertiary channel.

CHURN DRILL. A portable drilling machine arranged to successively raise and drop a heavy string of tools suspended from a drill line. By means of the successive blows the formation is chopped up and the hole deepened. The type of churn drill designed for placer sampling is often referred to as a "Keystone" drill or "placer" drill. A hand-powered type, used extensively in South America, is known as a "Ward" drill.

CLAIM. See – Mining claim.

CLEAN-UP. 1. The operation of collecting the gold or other valuable material from the recovery system of a dredge, hydraulic mine or other placer operation. 2. The valuable material resulting from a clean-up.

COARSE GOLD. The word "coarse", when applied to gold, is relative and is not uniformly applied. Some operators consider coarse gold to be that which remains on a 10-mesh screen. Others consider individual particles weighing 10 milligrams or more to be coarse gold. Some apply the term "coarse gold" to any particle that is relatively thick as compared to its diameter and can be easily picked up with the fingers.

COBBLE. A smoothly rounded stone, larger than a pebble and smaller than a boulder. (Fay)

COCOA MATTING. A heavy, coarse-woven fabric made of jute-like material and commonly placed on the bottom of a sluice to aid in saving fine gold.

COLLOIDAL GOLD. Gold in an extreme state of subdivision. In a true colloid, the individual particles are of almost molecular dimensions.

COLLUVIAL. Consisting of alluvium in part and also containing angular fragments of the original rocks. (Fay).

COLOR. A particle of metallic gold found in the prospector's pan after a sample of earth has been washed. Prospectors say, "The dirt gave me so many colors to the pan". (Fay)

CONCENTRATE. (verb) To separate a metal or mineral from its ore or from less valuable material. (noun) The product of concentration.

CONCENTRATION. The removal by mechanical means of the lighter and less valuable portions of ore. (Fay)

CONFLUENCE. A junction or flowing together of streams; the place where streams meet. (Fay)

CONGLOMERATE. Rounded waterworn fragments of rock of pebbles, cemented together by another mineral substance. (AGI)

CORE. See – Drill core.

CORE FACTOR. In churn drilling, when the casing is driven downward ahead of the drill bit, it should take in a cylinder of gravel having a diameter equal to the effective diameter (cutting edge) of the drive shoe. If the effective diameter of the shoe were the same as the inside diameter of the casing, a 1-foot drive would produce a 1-foot core core rise inside the casing. But this is not so. Take for example a standard 6-inch casing equipped with a new, 7½-inchdrive shoe. The effective area of the shoe is 44.17 square inches while that of the casing is about 26 square inches. As a result, when driven, the core should rise 44.17/26 = 1.7 or, in other words, there should be a 1.7-foot core rise inside the casing for each foot of drive. Here, the CORE FACTOR is 1.7. The core factor will, of course, vary according to the combination of casing and drive shoe used, and it will vary with the amount of wear on a given shoe. The core rise per foot of drive is less commonly referred to as the SHOE FACTOR, but to do so, invites risk of confusing it with other factors or terminology. See – Drive shoe factor; Pipe factor; Casing factor; Drill factor; Radford factor.

CORE RISE. The measured length of the cylinder of gravel entering a churn drill casing as it is driven downward. For example, a standard 6-inch casing fitted with a 7½-inch drive shoe should produce a core rise of 1.7 feet per foot of drive. The difference between the actual core rise and the theoretical rise is sometimes used as a factor for adjusting drill hole sample values.

CORRECTED VALUE. See – Adjusted value.

CRADLE. See – Rocker.

CREEK PLACER. Gravel deposits in the beds and intermediate flood plains of small streams. (Brooks)

CREVICING. A small-scale mining method in which the miner removes detrital material from cracks and crevices in the bedrock, usually by means of pry bars and long-handled spoons, and washes the material to recover its gold content.

CRIBBING. Close timbering, as the lining of a shaft. (Fay) In placer work, cribbing may be needed to support the walls of shaft or test pit put down in loose or wet ground.

DEBRIS. The tailings from hydraulic mines.

DEEP LEAD. (pronounced leed) A gold-bearing alluvial deposit buried below a considerable thickness of soil, lava or other barren material. See — Tertiary channel.

DESERT PLACER. See — Dry placer.

DETRITUS. A general name for incoherent sediments, produced by the wear and tear of rocks through the various geological agencies. The name is from the Latin for "worn" rock waste. (Fay) A deposit of such material.

DIP BOX. A modification of the sluice box used for small-scale mining where water is scarce. It generally consists of a short sluice made of 1 x 12-inch lumber, and standing on legs arranged to provide a steep slope. The gold-bearing material is washed in batches by first shoveling it into the upper end of the dip box and then pouring water over it, usually from a dipper.

DIRT. A miner's term for auriferous gravel or for the material being worked. See — Pay dirt.

DISCHARGE HEAD. The vertical distance from the center of a pump to the center of the discharge outlet where the water is delivered, to which must be added the loss due to friction of the water in the discharge pipe.

DISCOVERY. The finding of a valuable mineral deposit in place upon a mining claim. Although "discovery" and "valuable", as they relate to mining claims, have not been defined by statute, a long history of court decisions have held that in order for a location to be valid, there must be a discovery of mineral within the limits of the claim and the discovery must be such as would justify a person of ordinary prudence in the further expenditure of time and money, with reasonable prospect of success in developing a profitable mine. In some decisions the word "valuable" is interchanged with "profitable"

DISCOVERY CLAIM. (Alaska) A claim covering the initial discovery on a creek. Subsequent claims are commonly designated as one above, two above, three above; one below, two below, etc., depending on their position in relation to the discovery claim.

DOODLEBUG. 1. Miners' term for a dragline dredge. 2. A divining rod or similar device supposedly useful for locating gold or other valuable minerals. See — Dragline dredge.

DRAGLINE. A power shovel equipped with a long boom and a heavy digging bucket that is suspended from a hoisting line and is pulled toward the machine by means of a "drag" line. By manipulating the two lines (wire ropes), the bucket can be caused to dig, carry, or dump the excavated material. Such a machine is more properly called a dragline excavator. See — Dragline dredge.

DRAGLINE DREDGE. A dragline dredge consists of two units; a self-propelled power shovel equipped with a dragline bucket, and a floating washing plant which is similar to, but usually smaller than that of a bucket-line dredge. The washing unit contains a hopper for receiving gravel dug by the dragline; a

126

revolving screen; riffled sluices or other gold-saving equipment, and a tailings stacker. Dragline dredges are generally employed to mine relatively small, shallow deposits that are too small to amortize a bucket-line dredge.

DREDGE. A machine, operated by power, and usually mounted on a flat-bottomed hull provided with the equipment necessary to dig, process, and dispose of alluvial or other unconsolidated materials of a type found at the bottom of rivers or in certain terrestrial and offshore deposits. See – Bucklet-line dredge; Dragline dredge; Jet dredge; Suction dredge.

DREDGE SECTION. The depth of gravel, or a particular vertical section within a placer deposit, that will pay to mine by dredging.

DRIFT. (geol.) Any rock material, such as boulders, till, gravel, sand, or clay, transported by a glacier and deposited by or from the ice or by or in water derived from the melting of the ice. (Fay)

DRIFT. (mining) 1. A sub-tunnel running from the main tunnel to prospect for the pay lead; 2. A sub-tunnel run from the main tunnel across the pay lead to block out the ground and to facilitate its working; 3. Generally, a sub-tunnel. (Dunn, R. L.)

DRIFT MINING. A method of mining gold-bearing gravel by means of drifts, shafts or other underground openings, as distinguished from surface methods for placer mining.

DRILL. See – Churn drill.

DRILL CORE. A cylindrical core of sand and gravel forced upward into the drill casing as the casing or "drive pipe" is forced into the deposit, usually ahead of the drill bit. See – Core rise.

DRILL FACTOR. A figure used to designate the effective area of a drive shoe used in placer sampling. For example: A new, 7½-inch drive shoe has an open area of 0.306 sq. ft. but to allow for wear and other variables, some engineers use a lesser figure (commonly 0.27) in their value calculations. The figure so used is referred to as the DRILL FACTOR. See – Radford Factor; Core Factor; Volume Factor; Drive shoe factor.

DRILL LOG. The record of a drill hole, usually recorded on a prepared form as the work progresses. The usual placer log, in addition to showing the drilling progress, type of material penetrated, its mineral content, etc., will also show the type and size of equipment used, personnel employed, cause of delays, and other details of the work. A complete log will also show the essential calculations and all factors used in arriving at the reported value.

DRIVE PIPE. See – Casing.

DRIVE SHOE. A hardened steel protective shoe attached to the lower end of a drive pipe or casing. The drive shoe is usually slightly larger in diameter than the casing and is provided with a beveled cutting edge. See – Casing.

DRIVE SHOE FACTOR. The depth to which a churn drill casing must be driven to take in a sample volume of 1 cubic yard. For example, a standard 6-inch drive pipe equipped with a new, 7½-inch drive shoe would be driven 88 feet to cut out a theoretical sample volume of 1 cubic yard. This is less commonly called the PIPE FACTOR, or CASING FACTOR. See – Radford factor.

DRY DIGGINGS. In the 1850's, placers in or along the banks of California's rivers were known as "Wet diggings", and those in the dry ravines adjacent to the rivers were referred to as "Dry diggings". Compare with DRY PLACER.

DRY PLACERS. Placers in arid or semiarid regions, or generally where surface water is not available.

DRY WASHER. A device for recovering gold or other heavy minerals from dry alluvial material without the use of water. The typical dry washer is a small, hand-powered machine employing a sloping riffle board and a bellows or blower arrangement. The bottom of the riffle board is made of some porous material such as heavy cloth. Puffs of air forced up through the bottom by the bellows or blower, cause the lighter materials to hop over the riffles and work their way through the machine, while the gold or other heavy materials lodge behind the riffle bars.

DRY WASHING. The extraction of gold or other minerals from dry sand and gravel by the use of machines in which air is employed as a separating medium.

DRYLAND DREDGE. A mechanical washing plant, sometimes of appreciable size, designed to follow a dragline, or other excavator, as the mining cut advances. Some are equipped with trommel-type revolving screens and rock stackers, and are mounted on crawler-type tracks.

DUMP. 1. The fall immediately below a hydraulic mine outlet and in particular, the area available for tailings storage. 2. A specially prepared place outside of a drift mine, usually near the portal, where the pay gravel is deposited preparatory to washing. 3. A pile or heap of material, usually waste material, extracted from a mine.

DUST. See – Gold dust.

DUTY. 1. A measure of the effectiveness of water employed in hydraulic mining, usually expressed as the number of cubic yards of gravel washed per miners' inch per day (M.I.D.). The duty varies with the coarseness of gravel, height of bank, grade, available head, etc., usually varying from 1 to 7 cubic yards per miners' inch per 24 hours. 2. The effectiveness of water generally.

ELECTROSTATIC SEPARATOR. A device employing charged fields with little or no current flow, and used to extract or separate the component minerals of sands or heavy mineral concentrates. Speaking generally, electrostatic separators do not make sharp separations and they are sensitive to humidity, temperature, and other variables. Electrostatic separators have not found wide application in placer mining. Compare with High Tension separator.

ELEVATED SLUICE. See — Trestle sluice.

ELEVATOR. A device for ejecting gravel or tailings from a hydraulic mine pit. See — Hydraulic elevator; Rubel elevator.

ELUVIAL DEPOSIT. See — Eluvium.

ELUVIUM. Loose material resulting from decomposition of rock. Eluvial material may have slumped or washed downhill for a short distance but it has not been transported by a stream.

EMPIRE DRILL. See — Banka drill.

EOCENE. One of the earliest of the epochs into which the Tertiary period is divided; also the series of strata (and auriferous gravels) deposited at that time. Specifically, an epoch of the Tertiary between the Paleocene and Oligocene.

EROSION CYCLES. The Earth's erosional land forms develop in successive stages which can be divided into three broad categories — youth, maturity, and old age. Youthful land forms in the erosion cycle are featured by steep, narrow, V-shaped canyons and fast-cutting streams. In time, as the valleys deepen, they become wider and have gentler slopes. In early maturity they are roughly U-shaped instead of V-shaped. In late maturity downcutting has become slow, and conspicious flats develop. Old age is marked by wide flat valleys or peneplains, over which sluggish streams follow meandering courses. The mature stage is most favorable for the development of extensive placers.

EXPANDED METAL. (Expaned-metal lath) A type of punched-metal screen. The style commonly used in placer mining, for saving fine gold, consists of a latticework of diamond-shaped openings (about 3/4" x 1½") separated by raised metal strands that have a decided slope in one direction. When installed as riffles, with this slope leaning downstream, eddies form beneath the overhangs, thus creating conditions well-suited for the saving of fine gold. When used as riffles, expanded metal is generally placed over cocoa matting or similar material. A flat-lying style of expanded metal (without overhangs) is less-suited for this use.

FALSE BEDROCK. A hard or relatively tight formation within a placer deposit, at some distance above true bedrock, upon which gold concentrations are found. Clay, volcanic ash, caliche or "tight" gravel formations can serve as false bedrocks. A deposit may have gold concentrations on one or more false bedrocks, with or without a concentration on true bedrock.

FANNING CONCENTRATOR. See — pinched sluice.

FINE GOLD. 1. Pure gold, i.e., gold of 1000-fineness. 2. Gold occurring in small particles such as those which would pass a 20-mesh screen but remain on 40-mesh.

FINENESS. The proportion of pure gold in bullion or in a natural alloy, expressed in parts per thousand. Natural gold is not found in pure form; it contains varying proportions of silver, copper and other substances. For example, a piece of natural gold containing 150 parts of silver and 50 parts of copper per thousand, and the remainder pure gold, would be 800-fine. The average fineness of placer gold obtained in California is 885.

FINES. 1. The sand or other small-size components of a placer deposit. 2. The material passing through a screen during washing or other processing steps of a placer operation.

FIRE ASSAY. The assaying of metallic ores, usually gold and silver, by methods requiring furnace heat. (Fay) Fire assaying, in essence, is a miniature smelting process which recovers and reports the *total* gold content of the assay sample, including gold combined with other elements, or mechanically locked in the ore particles. Consequently, the gold value indicated by fire assay is not necessarily recoverable by placer methods. For this and other reaons, the gold content of placer material is not normally determined by fire assay. See – Free gold assay.

FLAKY GOLD. Very thin scales or pieces of gold.

FLASK. The unit of measurement for buying and selling mercury (quicksilver). A standard iron flask contains 76 pounds of mercury.

FLAT. An essentially level gravel bar or deposit along the banks of a river.

FLOAT. A term much used among miners and geologists for pieces of ore or rock which have fallen from veins or strata, or have been separated from the parent vein or strata by weathering agencies. Not usually applied to stream gravels. (Fay)

FLOAT-GOLD. Flour gold. Particles of gold so small and thin that they float on and are liable to be carried off by the water. (Fay) See – Flood gold.

FLOOD GOLD. Fine-size gold flakes carried or redistributed by flood waters and deposited on gravel bars as the flood waters recede. Flood gold sometimes forms superficial concentrations near the upstream end of accretion bars. See – Float gold.

FLOOD PLAIN. That portion of a river valley, adjacent to the river channel, which is built of sediments during the present regimen of the stream and which is covered with water when the river overflows its banks at flood stages. (AGI)

FLOTATION. The minimum working draft of a dredge. When a dredge "digs flotation" it excavates the ground to the minimum depth required for floating the dredge. This is usually done when passing through tailings or moving between nearby working areas.

FLOUR GOLD. The finest gold dust, much of which will float on water. (Fay) Flour gold, such as that found along the Snake River in Idaho, commonly runs 3 million colors to the ounce.

FLOURED MERCURY (QUICKSILVER). The finely granulated condition of quicksilver, produced to a greater or less extent by its agitation during the amalgamation process. The coating of quicksilver with what appears to be a thin film of some sulphide, so that when it is separated into globules these refuse to reunite. Also called Sickening and Flouring. (Fay)

FLUVIAL. Of, or pertaining to rivers; produced by river action, as a fluvial plain. (Fay)

FLUVIATILE. Caused or produced by the action of a river; fluvial. (Fay)

FLUVIOGLACIAL. Produced by streams which have their source in glacial ice. (Fay) See — Glaciofluvial.

FLUVIO-MARINE. Formed by the joint action of a river and the sea, as in the deposits at the mouths of rivers. (Fay)

FOOL'S GOLD. A substance which superficially resembles gold; usually pyrite, a sulphide of iron, FeS_2.

FREE GOLD. Gold uncombined with other substances. Placer gold. (Fay)

FREE GOLD ASSAY. A procedure carried out to determine the free gold content of an ore. In the case of placer material; a procedure to determine the amount of gold recoverable by gravity concentration and amalgamation.

FREE-WASH GRAVEL. Gravel that readily disintegrates and washes in a sluice. Loose, clay-free gravels such as those found in accretion bars are generally free-wash gravels.

GIANT. See — Hydraulic giant; also Intelligiant.

GLACIAL. Pertaining to, characteristic of, produced or deposited by, or derived from a glacier, (AGI)

GLACIOFLUVIAL. Of, pertaining to, produced by, or resulting from combined glacier action and river action. (Fay) See — Fluvioglacial.

GOLD DUST. A term once commonly applied to placer gold, particularly gold in the form of small colors.

GOLD PAN. See — pan.

GOLD-SAVING TABLE. The sluices used aboard a dredge are customarily called gold-saving tables, rather than sluice boxes.

GRADE. 1. The amount of fall or inclination from the horizontal in ditches, flumes, or sluices; usually measured in inches fall per foot of length or inches fall per section of sluice. 2. The slope of a land or bedrock surface; usually measured in percent. A one percent grade is equivalent to a rise or fall of one foot per hundred. 3. The slope of a stream, or the surface over which the water flows; usually measured in feet per mile. Streams having a grade of about 30 feet per mile favor the accumulation of placers, particularly where a fair balance between transportation and deposition is maintained for a long time. 4. The relative value or tenor of an ore, or of a mineral product.

GRADED STREAM. A stream in equilibrium, that is, a stream or a section of a stream that is essentially neither cutting or filling its channel.

GRAIN. A unit of weight equal to 0.0648 part of a gram, 0.04167 part of a pennyweight, or 0.002083 part of a troy ounce. There are 480 grains in a troy ounce. A grain of fine gold has a value of 7.29 cents (@$35/oz.).

GRAM. A unit of weight in the metric system equal to 15.432 grains, 0.643 pennyweight, or 0.03215 troy ounce. There are 31.003 grams in a troy ounce. A gram of fine gold has a value of $1.12 (@ $35/oz.).

GRAVEL. A comprehensive term applied to the water-worn mass of detrital material making up a placer deposit. Placer gravels are sometimes arbitrarily described as "fine" gravel, "heavy" (large) gravel, "boulder" gravel, etc.

GRAVEL MINE. A placer mine; a body of sand or gravel containing particles of gold. (Fay)

GRAVEL-PLAIN PLACERS. Placers found in gravel plains formed where a river canyon flattens and widens or more often, where it enters a wide, low-gradient valley.

GRIZZLY. An iron grating which serves as a heavy-duty screen to prevent large rocks or boulders from entering a sluice or other recovery equipment.

GROUND SLUICING. A mining method in which the gravel is excavated by water not under pressure. A natural or artificial water channel is used to start the operation and while a stream of water is directed through the channel or cut, the adjacent gravel banks are brought down by picking at the base of the bank and by directing the water flow as to undercut the bank and aid in its caving. Sluice boxes may or may not be used. Where not used, the gold is allowed to accumulate on the bedrock awaiting subsequent clean-up. A substantial water flow and adequate bedrock grade are necessary. See — Booming.

GUTTER. The lowest portion of an alluvial deposit; commonly a relatively narrow depression or trough in the bedrock. In some placers the pay streak is largely confined to a narrow streak or "gutter".

HAND DRILL. See — Ward drill. Also see — Banka drill and Empire drill.

HEAD. 1. A measure of (water) pressure. 2. The height of a column of water used for hydraulicking. For example, a hydraulic mine in which the point of water discharge is 200 vertical feet below the intake point (of the pipe) would be said to be working with a 200-foot head.

HEAVY GOLD. 1. Gold in compact pieces that appear to weigh heavy in proportion to their size. 2. Rounded, "shotty" or "nuggety" gold.

HEAVY MINERALS. The accessory detrital minerals of a sedimentary rock, of high specific gravity. (AGI) The black sand concentrate common to placers, would more properly be called a 'heavy-mineral' concentrate.

HIGH-GRADE. 1. Rich ore. 2. To steal or pilfer ore or gold, as from a mine by a miner. (Fay)

HIGH-GRADER. One who steals and sells, or otherwise disposes of high-grade or specimen ores. (Fay)

HIGH TENSION SEPARATOR. A machine, essentially consisting of a rotating drum, upon which a thin layer of dry sand or mineral grains are fed, and an electrode suspended above the rotating drum, or rotor. The electrode furnishes a high voltage discharge at high current flow. High tension separators employ a high rate of electrical discharge to separate various minerals according to their relative conductivity. Some are pinned to the rotor while others are attracted toward the electrode, with a resultant "lifting" effect. The pinning and lifting effects, imparted in varying degrees to different minerals, flattens or heightens their respective trajectories as they leave the rotor. Adjustable splitters placed in the trajectory are employed to cut selected minerals or groups of minerals from the thus stratified stream of material. High tension separators differ from electrostatic separators in that the latter employ charged fields with little or no current flow. High tension separators are extensively used for separating heavy minerals recovered from beach sands, monazite placers, etc.

HILLSIDE PLACERS. A group of gravel deposits intermediate between the creek and bench placers. Their bedrock is slightly above the creek bed, and the surface topography shows no indication of benching. (Brooks)

HORN SPOON. See – spoon.

HUMPHRYS SPIRAL. See – Spiral concentrator.

HYDRAULIC DREDGE. A dredge in which the material to be processed is excavated and elevated from the bottom of a stream or pond by means of a pump or a water-powered ejector. Large hydraulic dredges may be equipped with a digging ladder which carries the suction pipe and a motor-driven cutter head, arranged to chop-up or otherwise loosen material directly in front of the intake pipe. Dredges having this configuration employ a deck-mounted suction pump and they may carry the mineral recovery equipment on board the dredge or more commoly, they may transport the excavated material, by means of a pipe line, to a recovery plant mounted on independent barges or on the shore. See – Jet dredge; also bucket-line dredge.

HYDRAULIC ELEVATOR. A near-vertical pipe employed in hydraulic mining to raise excavated material from the working place to an elevated sluice, or to a disposal area, by means of a high-pressure water jet inducing a strong upward current in the elevator pipe. see – Rubel elevator.

133

HYDRAULIC GIANT. The nozzle assembly used in hydraulic mining. The giant is provided with a swivel enabling it to be swung in a horizontal plane, and it may be elevated or depressed in a vertical plane. Nozzle sizes range from 1 to 10 inches in diameter and the largers sizes are provided with a deflector, enabling them to be moved with little effort. In California, giants discharging as much as 15,000 gallons per minute in a single stream at a nozzle pressure of over 200 pounds per square inch, have been used. The giant is also known as a "Monitor". Both terms stem from manufacturer's trade names. See – Intelligiant.

HYDRAULIC MINING. A method of mining in which a bank of gold-bearing earth or gravel is washed away by a powerful jet of water and carried into sluices, where the gold separates from the earth by its specific gravity. (Fay)

HYDRAULIC MONITOR. See – Hydraulic giant.

HYDRAULICKING. Mining by the hydraulic method. Note spelling.

INCHES OF WATER. A common expression denoting the quantity of water (in miners' inches) available or being used in a placer operation. See – Miners' inch.

INDICATED VALUE. The value of a placer sample, derived by formula, before making adjustments to compensate for excess or deficient core rise, in the case of churn drilling; or before applying shaft factors, boulder factors, or other empirical corrections. See – Adjusted value.

INLET. The point where a channel is cut off by a ravine or canyon on the upstream end. (Dunn, R.L.) Usually applied to buried Teritary channels. Compare with Breakout; and with Outlet.

INTELLIGIANT. The trade name for a hydraulic giant that is provided with water-powered piston and cylinder arrangements to control its vertical and horizontal traverses. Some models can be rigged for automatic operation and can run unattended in a preset arc or pattern. See – Hydraulic giant.

IRON SAND. 1. Magnetite or ilmenite-rich sand. 2. Black sand concentrate containing an abundance of magnetite.

JET DREDGE. A form of hydraulic dredge. Jet dredging equipment may range from a simple, self-contained pipe-like venturi containing riffles, that is carried by a diver and operates entirely underwater to larger and more elaborate surface units carried on inflated rubber tubes or styrofoam floats. These devices, operated by one or two men, are similar in two ways: 1. They rely on a water jet and venturi effect to pick up unconsolidated stream-bottom materials and carry them to a gold recovery device, usually riffles. 2. The suction intake is normally hand-held and is guided by a diver working on the stream bottom. The typical jet "dredge" entails a small or modest capital outlay and is typically used for recreation-type mining. See – Hydraulic dredge.

JET DRILL. A churn-type drill employing a string of reciprocal hollow rods equipped with a drill bit. Water is pumped through the rods and discharged through an orifice near the bit. Cuttings resulting from the chopping action of the bit are carried to the surface by wash water rising between the drill rods and casing. Rods are added as the hole deepens, thus the drill cable does not go down the hole as would be the case in conventional churn drilling. Jet drills are well suited to sampling low-value minerals, such as ilmenite, occurring in beach deposits.

JIG. A machine in which heavy minerals are separated from sand or gangue minerals on a screen in water, by imparting a reciprocating motion to the screen or by the pulsation of water through the screen. Where the heavy mineral is larger than the screen openings, a concentrate bed will form on top of the screen. Where the heavy mineral particles are smaller than the screen openings, a fine-size concentrate will be collected in a hutch beneath the screen.

KEYSTONE CONSTANT. See – Radford factor.

KEYSTONE DRILL. See –Churn drill.

KNUDSEN BOWL. See – Ainlay bowl.

LACUSTRINE DEPOSITS. Deposits formed in the bottom of lakes. (Fay) Compare with Lake-bed placers.

LAKE-BED PLACERS. Placers accumulated in the beds of present or ancient lakes that were generally formed by landslides or glacial damming. (Brooks) It should be noted that a lake-bed (or lake-bottom) placer might actually be a drowned stream placer.

LAVA. The term 'Lava' as used by a placer miner, may designate any solidified volcanic rock including volcanic agglomerates.

LEAD. (pronounced leed) Deeply buried placer gravel, where rich enough to work, and particularly when in a well-defined bed, is often termed the "lead" or, "pay lead".

LIGHT GOLD. Gold that is in very thin scales or flakes or in pieces that look large as compared to their weight. See – Flood gold.

LITTORAL. Pertaining to the shore of a lake, sea, or ocean.

LOCATION. See – Mining claim.

LONG TOM. 1. A small, sluice-type gold washer widely used in California during the 1850's and 60's. The early long tom was built in two sections; a washing box equipped with a perforated plate to screen out the rocks; followed by a short sluice containing riffles. 2. A short auxiliary sluice used aboard a dredge to further reduce concentrate taken from the dredge riffles at clean-up time. 3. A short sluice used to wash placer samples.

LOW-GRADE. A term applied to ores relatively poor in the metal for which they are mined; lean ore. (Fay)

MAGNETIC SEPARATOR. A device in which a strong magnetic field is employed to remove magnetic materials from a sand or a concentrate, or to selectively remove or separate their constituent minerals. Magnetic separators are commonly used in conjuction with high tension separators to process the heavy mineral concentrates obtained from beach sands, monazite placers, tin placers, etc.

MARINE MINING. The exploitation of sea-bottom mineral deposits, including placers. See — Marine placer.

MARINE PLACER. A deposit of placer-type minerals on the ocean or sea bottom beyond the low-tide line, as distinguished from beach placers. Some marine placers may contain material related to beach deposits formed during periods of low sea level. Others may contain stream-type placers or mineral concentrations formed on land and later drowned by a lowering of the coastal region.

MATURITY (MATURE VALLEY). See — Erosion cycles.

MEANDER. One of a series of somewhat regular and looplike bends in the course of a stream, developed when the stream is flowing at grade, through lateral shifting of its course toward the convex sides of the original curves. (Fay)

MEDIUM-SIZE GOLD. Gold of an approximate size that will pass through a 10-mesh screen and remain on a 20-mesh screen. Compare with Coarse gold; also Fine gold.

MERCURY. A heavy, silver-white liquid metallic element, useful in placer mining where its chemical affinity for gold is taken advantage of to help detain gold in a sluice box. Mercury placed in the riffles forms a gold amalgam which is removed at the time of clean-up and then retorted to recover the gold. The miners' term for mercury is "Quicksilver" or simply, "Quick". Symbol, Hg; specific gravity, 13.54.

MILLIGRAM. The one-thousandth part of a gram. As a matter of convenience, the milligram is widely used as the unit for reporting gold weights in placer samples. There are 31,103 milligrams in a troy ounce. With gold at $35 per troy ounce, 1 milligram of fine gold is worth 0.112 cent and 1 milligram of ordinary placer gold is worth about 0.1 cent, or in other words, 10 milligrams to the cent.

MINERS' INCH. A unit of water measurement. Originally it represented the quantity of water that will escape from an aperture one inch square through a two-inch plank, with a steady flow of water standing six inches above the top of the escape aperture. The miners' inch is now defined by statute in various states.

136

1 second-foot = 40 miners' inches in Arizona, California, Montana and Oregon.

= 50 miners' inches in Idaho, Nevada, New Mexico and Utah.

= 38.4 miners' inches in Colorado.

1 miners' inch equals 11.25 gallons per minute when equivalent to 1/40 second-foot.

1 miners' inch eauals 9 gallons per minute when equivalent to 1/50 second-foot.

MINERS' PAN. See – Pan

MINERS' SPOON. See – Spoon.

MINING CLAIM. That portion of the public mineral lands which a miner, for mining purposes, takes and holds in accordance with the mining laws. (Fay) A mining claim may be validly located and held only after the discovery of a valuable mineral deposit. See – Discovery.

MONAZITE. A phosphate of the cerium metals (cerium, didymium, lanthanum) and other rare-earth metals. Monazite-bearing alluviums have been mined for their thorium content (by dredging) in Idaho and elsewhere.

MONITOR. See – Hydraulic giant.

MORAINE. An accumulation of earth, stones, boulders, etc., carried and finally deposited by a glacier. A Moraine formed at the lower extremity of a glacier is called a TERMINAL Moraine; at the side, a LATERAL Moraine; in the center and parallel with its sides, a MEDIAL Moraine and beneath the ice but back from its end or edge, A GROUND Moraine. (Fay) Placer gold is found in some glacial Moraines and deposits of reworked Morainal material, that is, material reworked by streams; some have been dredged and worked by other placer methods.

MOSS MINING. (Mossing). The gathering of moss from the banks of gold-bearing streams for the purpose of burning or washing it, to recover its gold content. Under certain conditions, moss or similar vegetation will capture and hold small particles of gold being carried downstream by flood waters. See – Flood gold.

MUCK. (Alaska) A permanently frozen overburden overlying placer gravels in the interior of Alaska. It is composed of fine mud, organic matter and small amounts of volcanic ash. It varies in depth (thickness) from seldom less than 10 feet to 100 feet or more in places. This overburden (muck) must be removed and the underlying gravels thawed before dredging is possible.

NATIVE GOLD. 1. Metallic gold found naturally in that state. 2. Placer gold.

NUGGET. 1. A water-worn piece of native gold. The term is restricted to pieces of some size, not mere 'colors' or minute particles. Fragments and lumps of vein gold are not called 'nuggets', for the idea of alluvial origin is implicit. (Fay) 2. Anything larger than, say, one penny-weight or one gram may be considered a nugget. See – Pepita.

NUGGETY. Like or resembling a nugget; occurring in nuggets; also abounding in nuggets. (Fay)

OLD AGE. See — Erosion cycles.

OUTCROP. The exposure of bedrock or strata projecting through the overlying cover of detritus and soil. (AGI)

OUTLET. The point where a channel is cut off by a ravine or canyon on the downstream end. (Dunn, R.L.) Usually applied to buried Tertiary channels. Compare with Breakout; and with Inlet.

OVERBURDEN. Worthless or low-grade surface material covering a body of useful mineral. The frozen muck covering dredge gravels in Central Alaska is an example of placer overburden.

OFFSHORE DEPOSITS. Mineral deposits on the ocean or sea bottom beyond the low-tide line. See — Marine placer.

PAN. 1. A shallow, sheet-iron vessel with sloping sides and a flat bottom, used for washing auriferous gravel or other materials containing heavy minerals. It is usually referred to as a 'Gold pan', but is more properly called a 'Miners'pan'. Pans are made in a variety of sizes but the size generally referred to as "standard" has a diameter of 16 inches at the top, 10 inches at the bottom, and a depth of 2½ inches. Pans made of copper, or provided with a copper bottom are sometimes used for amalgamating gold. 2. (verb) To wash earth, gravel, or other material in a pan to recover gold or other heavy minerals.

PAN FACTOR. The number of pans of gravel equivalent to a cubic yard in place. Pan factors vary according to the size and shape of the pan, the amount of heaping when filling the pan, the swell of ground when excavated, and other factors. In practice, factors for a 16-inch pan range from 150 to 200; a factor of 180 is widely used.

PANNING. Washing gravel or other material in a Miners' pan to recover gold or other heavy minerals.

PATENT. A document by which the Federal Government conveys title to a mining claim. To obtain a mineral patent, the applicant must among other things, (1) make a valid mineral discovery (2) invest $500 in improvements, (3) pay for a boundary survey if lode minerals are applied for, (4) pay $2.50 per acre for the lands in a placer application, or $5.00 per acre for the lands in a lode application. See — Discovery.

PAY DIRT. Auriferous gravel rich enough to pay for washing or working. (Fay)

PAY LEAD. (pronounced leed) Where gravel is found rich enough to work, and if there is a well-defined bed of it, it is often termed the "pay lead" or, "lead". Compare with Pay streak.

PAY STREAK. A limited horizon within a placer deposit, containing a concentration of values or made up of material rich enough to mine. Pay streaks in gold placers are commonly found as more or less well-defined areas on or near bedrock and are commonly narrow, sinuous, and discontinuous. Compare with Pay lead.

PEDIMENT. Gently inclined planate erosion surfaces carved in bedrock and generally veneered with fluvial gravels. They occur between mountain fronts and valley bottoms and commonly form extensive bedrock surfaces over which the erosion products from the retreating mountain fronts are transported to the basins. (AGI)

PENEPLAIN. A land surface worn down by erosion to a nearly flat or broadly undulating plain; the penultimate stage of old age of the land produced by the forces of erosion. (AGI)

PENNYWEIGHT. A unit of weight equal to 24 grains, 0.05 troy ounce or 1.5552 grams. A pennyweight of fine gold has a value of $1.03, with gold at $20.67 per ounce. A pennyweight of fine gold has a value of $1.75, with gold at $35.00 per ounce.

PEPITA. (Spanish) A nugget; usually a smaller size.

PERMAFROST. Permanently frozen ground in Alaska, up to 100 or more feet in thickness. See – Muck.

PILOT SLUICE. A small, auxiliary sluice operated intermittently aboard a dredge to determine the amount of gold being recovered by the dredge during a given interval of time, or from a particular gravel section. The ratio of pilot sluice recovery to dredge recovery is determined for each dredge by empirical means.

PINCHED SLUICE. A film-type gravity concentrator employing a wedge-shaped trough, tapering to a narrow vertical opening at its discharge end. In use, heavy-gravity minerals migrate toward the bottom and are removed from the stratified discharge stream by means of splitters. Pinched sluice-type concentrators are used to remove heavy minerals, such as rutile and ilmenite, from beach sands. The CANNON CONCENTRATOR and FANNING CONCENTRATOR are of this type.

PIPE CLAY. Miners' term for clays or clay-like materials found in finely-laminated beds within the Tertiary gravels of California's Sierra Nevada region. Some may consist of volcanic material which has fallen into water, in the form of ash, and taken on a stratified form resembling clay in appearance.

PIPE FACTOR. The depth to which a churn drill casing must be driven to take in a sample volume of 1 cubic yard. For example, a standard 6-inch drive pipe equipped with a new, 7½-inch drive shoe would be driven 88 feet to cut out a theoretical volume of 1 cubic yard. This is sometimes called the CASING FACTOR but it is most commonly known as the DRIVE SHOE FACTOR. See – Radford factor.

PIPER. The man operating a hydraulic giant and directing its stream.

PIPING. Washing gravel with a hydraulic giant.

PITCH. Used in connection with the bedrock in the channel or rim to express decent. (Dunn, R. L.)

PITTING. The act of digging or sinking a pit, as for sampling alluvial deposits. (Fay)

PLACER. A place where gold is obtained by washing; an alluvial or glacial deposit, as of sand or gravel, containing particles of gold or other valuable mineral. In the United States mining law, mineral deposits, not veins in place, are treated as placers, so far as locating, holding, and patenting are concerned. (Fay) The term "placer" applies to ancient (Tertiary) gravels as well as to recent deposits, and to underground (drift mines) as well as to surface deposits.

PLACER DEPOSIT. A mass of gravel, sand, or similar material resulting from the crumbling and erosion of solid rocks and containing particles or nuggets of gold, platinum, tin, or other valuable minerals, that have been derived from the rocks or veins. (Fay)

PLACER MINING. That form of mining in which the surficial detritus is washed for gold or other valuable minerals. When water under pressure is employed to break down the gravel, the term HYDRAULIC MINING is generally employed. There are deposits of detrital material containing gold which lie too deep to be profitably extracted by surface mining, and which must be worked by drifting beneath the overlying barren material. To the operations necessary to extract such auriferous material the term DRIFT MINING is applied. (Fay)

POINT BAR. See – Skim bar.

POINTS. See – Thaw points.

PROSPECT DRILL. See – Churn drill.

PROSPECTING. 1. Used to qualify work merely intended to discover a pay lead in a drift mine, or to locate the channel. (Dunn, R. L.) 2. (generally) Searching for new deposits. 3. Drilling a known placer deposit to determine its value or delineate a minable area.

QUATERNARY GRAVELS. Gravels deposited from the end of the Tertiary, to and including the present time.

QUICKSILVER (or 'Quick'). See – Mercury.

RADFORD FACTOR. An arbitrary factor used by some engineers in the calculation of drill hole volumes and in turn, the drill hole values. This factor is based on an assumption that due to wear, etc., a 7½-inch drive shoe will take in 0.27 cu. ft. of core per foot of drive instead of the theoretical 0.306 cu. ft. In other words, it assumes a core volume of 1/100 cubic yard per foot of drive. Using this factor, the equation for calculating the drill hole value becomes: value of recovered gold in cents times 100, divided by depth of hole in feet, equals cents per cubic yard. Use of the Radford factor will upgrade the theoretical value by about 12 percent. The Radford factor is sometimes called the KEYSTONE CONSTANT. When reviewing drill logs or reports, care should be taken to determine the factor used. See — Drill factor.

RADIOACTIVE BLACKS. A group of dark colored, heavy minerals recovered by placer mining methods in Idaho and elsewhere, and valuable for their contained uranium, thorium, or rare earth components. They include such minerals as brannerite, euxenite, davidite, betafite, and samarskite. See — Rare Earth minerals.

RARE EARTH MINERALS. A group of widely distributed but relatively scarce minerals containing rare earth compounds, usually in combination with uranium, thorium, and other elements. Monazite and other rare earth minerals are obtained from placers in Idaho, and elsewhere. See — Monazite; also Radioactive blacks.

R/E. See — Recovery.

RECOVERY. 1. The amount or value of mineral recovered from a unit volume; in the case of gold placers, expressed as cents per cubic yard. 2. The amount of mineral extracted, expressed as a percentage of the total mineral content. 3. In gold dredging, the expression "R over E" (designated R/E) is used to compare actual recovery to expected recovery where R represents the actual returns and E represents the estimated recoverable value, after allowing for known or expected mining and metallurgical losses, etc. When recovery exceeds the initial estimate, the R/E will be shown as something greater than 100%, such as 105%, 110%, etc.

REPRESENTATIVE SAMPLE. See — Sampling.

RESIDUAL PLACER. Essentially, an in situ enrichment of gold or other heavy mineral, caused by weathering and subsequent removal of the lode or other parent material, leaving the heavier, valuable mineral in a somewhat concentrated state. In some cases, a residual placer may be essentially an area of bedrock, containing numerous gold-bearing veinlets that have disintigrated by weathering to produce a detrital mantle rich enough to mine. In some parts of California, such areas are known as SEAM DIGGINGS.

RETORT. A vessel with a long neck used for distilling the quicksilver from amalgam. (Fay)

RIFFLE. 1. The lining of a bottom of a sluice, made of blocks or slats of wood, or stones, arranged in such a manner that chinks are left between them. The whole arragement at the bottom of the sluice is usually called THE RIFFLES. In smaller gold-saving machines, as the rocker, the slats of wood nailed across the bottom are called RIFFLE-BARS, or simply RIFFLES. (Fay) 2. A groove in the bottom of an inclined trough or sluice, for arresting gold contained in sands or gravels. (Fay) 3. A shallow extending across the bed of a stream; a rapid of comparatively little fall. (AGI)

RIM ROCK (or RIM). The bedrock rising to form the boundary of a placer or gravel deposit. (Fay)

RIVER-BAR PLACERS. Placers on gravel flats in or adjacent to the beds of large streams. (Brooks)

RIVER MINING. The mining of part or all of a river bed after by-passing the stream by means of flumes or tunnels; or by use of wing dams to divert the river from the working area.

ROCKER. A short, sluice-like trough fitted with transverse curved supports, permitting it to be rocked from side to side, and provided with a shallow hopper at its upper end. The hopper bottom consists of a punched-metal plate containing holes about ¼-inch or ½-inch diameter. This is for the purpose of holding back the larger rocks which when washed, are discarded. A flow of water, aided by the rocking motion, carries the fine material down the trough where the gold or other heavy minerals are caught by riffles. Rockers are generally operated by hand but large, power-driven rockers are sometimes employed. When washing churn drill samples, rockers are often used without riffles, the recovery being made on the smooth wooden bottom much in the manner of panning. CRADLE is an obsolete term for rocker.

ROCKING. The process of washing sand or gravel in a rocker.

ROUGH GOLD. Gold that has not been appreciably worn or smoothed by movement and abrasion. It may be more angular than rounded and may have included or attached quartz particles. As a rule, rough gold is found near its place of origin.

RUBEL ELEVATOR. (pronounced Roo-bull) A form of elevator used in hydraulic mines, particularly those having insufficient bedrock grade for effective tailings disposal. It is essentially a large, inclined flume, through which gravel or tailings are driven by a strong water jet furnished by a hydraulic giant. A grizzly arrangement removes the fines for treatment in conventional sluices while the rocks are discharged from the upper end.

RUSTY GOLD. Free gold, that does not readily amalgamate, the particles being covered with a silicious film, thin coating of oxide of iron, etc. (Fay)

SALTING. 1. Intentional salting. The surreptitious placing of gold or other valuable material in a working place or in a sample to make it appear rich in mineral. It is done with intent to defraud. 2. Unintentional or innocent salting. The unintentional or accidental enrichment of a sample through erroneous procedure or carelessness, without intent to defraud.

SAMPLE. A portion of the ore systematically taken, by which its quality is to be judged. (Fay)

SAMPLING. Cutting a representative part of an ore deposit, which should truly represent its average value. Honest sampling requires good judgement and practical experience. (Fay) Parenthetically, it should be noted that in the case of gold placers, the high unit-value of gold, its extreme dilution within the gravel mass and its typically erratic distribution are factors which individually or combined, make it virtually impossible to obtain a truly representative sample. To this extent, the usual definitions of sampling do not apply to gold placers.

SAND PUMP. A special plunger-type vacuum pump used to remove the chopped-up drill core from a churn drill hole.

SAUERMAN EXCAVATOR. See — Slackline scraper.

SCALY GOLD. Small, rounded, flattened gold particles; usually quite thin in proportion to their diameter.

SCHIST. A crystalline rock that can be readily split or cleaved because of having a foliated or parallel structure. (Fay) Schist bedrocks, because of their rough, platy structure, generally make excellent gold catchers.

SCHISTOSE. Characteristic of, resembling, pertaining to, or having the nature of schist. (Fay)

SEA-BEACH PLACERS. Placers reconcentrated from the coastal-plain gravels by the waves along the seashore. (Brooks)

SEAM DIGGINGS. (California) Residual deposits consiting of decomposed bedrock filled with irregular swams of quartz containing gold. In California, seam diggings have been worked by the hydraulic method.

SECOND-FOOT. A unit of water measurement equivalent to one cubic foot per second or 448.83 gallons per minute. Commonly used to report the flow of streams.

SELF-SHOOTER. See — Booming.

SHAFT FACTOR. A correction factor applied to drill-hole values after a shaft has been sunk over the drill hole. The factor is based on the difference in values obtained from the drill hole and from the shaft; the shaft value generally being considered the more reliable of the two.

SHINGLE. 1. The flatter pebbles and cobbles in a stream deposit will often come to rest with their uppermost edge leaning slightly down-steam. This 'shingling' effect is used by placer miners to determine the direction of flow of ancient streams and it can be particularly useful when working drift mines. 2. Beach gravel, especially if consisting of flat or flattish pebbles.

SHOE FACTOR. See — Drive Shoe factor.

SHOTTY GOLD. Small granular pieces of gold resembling shot. (Fay) Any small, more or less rounded gold particle that is somewhat equidimensional rather than platy.

SICK MERCURY. See — Floured mercury.

SKIM BAR. An area near the upstream end of an accretion bar from which superficial concentrations of flood gold are mined by 'skimming' off a thin layer of gravel. They are sometimes known as POINT BARS, probably because of their proximity to the upper point of the accretion bar. See — Flood gold; also Accretion bar.

SKIN DIVING. The use of wet-type diving suits, with or without self-contained underwater breathing apparatus. Skin diving gear is generally used by the operators of small hydraulic dredges and by divers who search for underwater bedrock crevices from which gold-bearing materials may be retreived. See — Jet dredge.

SLACKLINE SCRAPER. Consists essentially of a head tower and a movable tail tower or tail block, supporting a track cable. A bucket or scraper running along the track cable can be raised and lowered by tightening or slackening the track cable. The digging bucket or scraper runs out by gravity and is pulled in by a drag cable. The hoisting machinery and in some cases a screening or washing plant are incorporated in the head tower. This arrangement is also known as a CABLEWAY SCRAPER. The SAUERMAN EXCAVATOR is of this type.

SLATE. A fine-grained rock formed by the compression of clay, shale, etc., that tends to split along parallel cleavage planes and to form a rough, platy bedrock, well suited for the retention of placer gold.

SLICKENS. A word sometimes used to designate the finer-size tailings, or mud, discharged from a placer mine. Sometimes synonymous with Slime.

SLUDGE. The fluid mixture of chopped up core and water that results from the drilling action in a churn drill hole. When the sludge is pumped from the hole, it becomes the sample for the particular section of hole that produced it.

SLUICE BOX. An elongated wooden or metal trough, equipped with riffles, through which alluvial material is washed to recover its gold or other heavy minerals. Small sluice boxes are commonly, but erroneously, called "Long Toms".

SLUICEPLATE. A shallow, flat-bottomed steel hopper arrangement at the head-end of a sluice box. A bulldozer is generally used to push gold-bearing gravel onto the sluiceplate, from where it is washed into the sluice by water issuing from a large pipe or by means of a small hydraulic giant.

SNIPER. An individual miner, usually a transient, who gleans a living from gravel remnants not worth working except by someone content with very modest gains. He usually works with simple hand tools and washes his gravel in a short sluice or dip box. Being transient and generally innocuous, he seldom owns or leases the land he works.

SODIUM AMALGAM. Mercury that has been treated with small amounts of metallic sodium to increase its affinity for gold and other metals.

SPECIFIC GRAVITY. The specific gravity of a substance is its weight as compared with the weight of an equal bulk of pure water. For example, placer gold with a specific gravity of about 19 is 19 times heavier than water.

The specific gravity of a mineral largely determines its susceptibility to recovery in simple gravity concentrators such as sluice boxes.

SPECIMEN GOLD. Nuggety gold or other forms suitable for the manufacture of natural-gold jewelry or for display purposes.

SPIRAL CONCENTRATOR. A wet-type gravity concentrator in which a sand-water mixture, flowing down a long, spiral shaped launder, separates into concentrate and tailings fractions. The concentrates are taken off through ports while the tailings flow to waste at the bottom. The HUMPHREYS SPIRAL, which employs this principle, is widely used for recovering heavy minerals from beach sands.

SPONGE. The somewhat porous mass of gold remaining after the mercury has been removed from a gold amalgam by heating.

SPOON. A shallow, oblong vessel, at one time made from a section of ox horn but now made of metal. Used to test small samples of gold-bearing material by washing, in a manner similar to panning. More properly called a MINERS' SPOON or, HORN SPOON.

SPOTTED GRAVEL. Where gold is erratically distributed through a deposit, the term "spotted" or, "spotty"gravel is sometimes applied to it.

STRIP. To remove the overlying earth, low-grade, or barren material from a placer deposit.

STRUCK CAPACITY. Level-full, that is, the capacity of a container filled even with its rim or top.

SUBMARINE PLACER. See – Marine placer.

SUCKER. 1. A syringe used to remove material from underwater crevices in the bedrock. 2. A small, hand-held jet dredge of the type carried underwater.

SUCTION DREDGE. See – Hydraulic dredge; also Jet dredge.

SUCTION LIFT. The vertical distance from the level of the water supply to the center of a pump, to which must be added the loss due to friction of the water in the suction pipe.

SURF WASHER. A small sluice, somewhat similar to a long tom; used to recover gold from beach sands. The surf washer is placed so the incoming surf rushes up the sluice, washing material from a hopper and upon retreating carries it over the riffles.

SWELL. The expansion or increase in volume of earth or gravel upon loosening or removal from the ground. The average swell of gravel is around 25% and sometimes as high as 50%.

TAIL. (verb) Manipulating the concentrate product in a gold pan in such a way that the heavier minerals and in particular the gold colors string out in the bottom of the pan in a long, narrow "tail", where they can be readily inspected or counted. This is referred to as "tailing a pan."

TAILINGS. The washed material which issues from the end of a sluice or other recovery device in a placer operation. The tailings from hydraulic mines are generally referred to as "debris" in legislative documents.

TENOR. The percentage or average metallic content of an ore. (Fay) As commonly used, it is synonymous with an approximate, or a general value rather than a precisely known value.

TERRACE. A relatively flat, and sometimes long and narrow surface, commonly bounded by steep upslopes and downslopes on opposite sides. Gravel terraces may be stepped, and they are commonly dissected by transverse drainage patterns.

TERTIARY. The earlier of the two geologic periods comprised in the Cenozoic era, in the classification generally used. Also, the system of strata deposited during that period. (AGI)

ERA	PERIOD		YEARS AGO
CENOZOIC	QUATERNARY	RECENT	
		PLEISTOCENE	
			1,000,000
	TERTIARY	Pliocene	
			11,000,000
		Miocene	
			30,000,000
		Oligocene	
			40,000,000
		Eocene	
			58,000,000
		Paleocene	
			60,000,000
MESOZOIC		CRETACEOUS	

TERTIARY CHANNELS. Ancient gravel deposits, often auriferous, composed of Tertiary stream alluvium. Tertiary gravels are abundant in the Sierra Nevada gold belt of California where many have been covered by extensive volcanic eruptions and subsequently elevated by mountain uplifts, and are now found as deeply-buried channels, high above the present stream beds.

TEST PIT. See – PITTING.

THAW POINTS. Water pipes driven into frozen gravel, through which water at natural temperature is ciculated for weeks or months, to thaw the ground ahead of dredging. Where used in Alaska, points are usually spaced 16 feet apart. Once thawed, the ground does not freeze again and thawing is usually carried one or two seasons ahead of the dredge.

TIGHT GRAVEL. A hard, or compact gravel that is not cemented, but requires something more than normal effort to excavate. Compare with CEMENTED GRAVEL.

TILL. Nonsorted, nonstratified sediment carried or deposited by a glacier (AGI)

TOP WASH. A deposit of gravel, not in a channel on the bedrock, but resting on cement overlying the bottom deposit. (Dunn, R. L.)

TRACE. A very small quantity of gold; usually a speck too small to weigh. In reporting samples it is abbreviated tr.

TRESTLE SLUICE. A moveable steel sluice constructed on a skid or track-mounted trestle; usually provided with a hopper, grizzly and wash water system, and fed by a dragline or similar excavator. Also called an ELEVATED SLUICE.

TROMMEL. A heavy-duty revolving screen used for washing and removing the rocks or cobbles from placer material prior to treatment in the sluices, gold-saving tables, or other recovery equipment.

TROY OUNCE. The one-twelfth part of a pound of 5760 grains; that is 480 grains. It equals 20 pennyweights, 1.09714 avoidupois ounces, 31.1035 grams, or 31,103 milligrams. This is the ounce designated in all assay returns for gold, silver or other precious metals. (Fay)

TUNNEL. The nearly horizontal excavated opening from the surface into the mine. (Dunn, R. L.)

UNDERCURRENT. A large, flat, broad, branch sluice, placed beside and a little lower than the main sluice. This apparatus is riffled like the sluice, but being much wider than the latter, allows the water to spread out in a thin sheet over its surface, thereby so abating the velocity of the current that the very fine gold, including the rusty particles, is more apt to be caught here than in the sluice. (Fay) Undercurrents are usually fed with fine-size material taken from the main sluice by means of a grizzly placed in the sluice bottom, near the discharge end.

UPPER LEAD. (pronounced leed) A pay lead in a top wash or in the gravel deposit considerably above the bedrock. (Dunn, R. L.)

VALUES. The valuable ingredients to be obtained, by treatment, from any mass or compound; specifically, the precious metals contained in rock, gravel, or the like. (Fay)

VOLUME FACTOR. The volume of sample which should be taken into a churn drill casing for each foot of drive. For example, a standard 6-inch drive pipe equipped with a new, 7½-inch drive shoe will theoretically take in a volume of 530 cubic inches, or, 0.306 cubic foot per foot of drive. See — Core factor; Drive shoe factor; Drill factor.

WARD DRILL. A lightweight, hand-powered churn drill widely used in South America, particularly in remote areas where access is difficult and manpower is cheap. The drilling tools are suspended from a tripod and the reciprocating motion provided by a simple spudding arm known as a "Diablo". Sometimes referred to as a HAND DRILL.

WASH. 1. A Western miners' term for any loose, surface deposits of sand, gravel, boulders, etc. 2. The dry bed of an intermittent stream, sometimes at the bottom of a canyon. Also called Dry wash. 3. To subject gravel, etc. to the action of water to separate the valuable material from the worthless or less valuable; as to wash gold. (Fay) In drift mining (California), the term "Wash" is used indifferently in describing channel gravel, volcanic mud flows, or masses of lava boulders. (Dunn, R. L.)

WASTE. Valueless material such as barren gravel or overburden. Material too poor to pay for washing.

WATER TABLE. The upper limit of the portion of the ground wholly saturated with water. This may be very near the surface or many feet below it. (Fay)

WEATHERING. The group of processes, such as the chemical action of air and rain water and of plants and bacteria and the mechanical action of changes of temperature, whereby rocks on exposure to the weather change in character, decay, and finally crumble into soil. (Fay)

WING DAM. A dam built partially across a river to deflect the water from its course. (Fay) See — River mining.

WING FENCE. A V-shaped wall, usually made of heavy timber and attached to the head of a sluice and arranged to guide gravel into the sluice as it is swept from the pit by a hydraulic giant.

YARDAGE. 1. The number of cubic yards of gravel mined or put through a washing plant in a shift or a day. 2. A measured block of gravel.

YIELD. The quantity or gross value of minerals extracted from a deposit.

YOUTH. See — Erosion cycles.

ZIRCON. A mineral of widespread occurrence as small crystals in igneous rocks. Composition, zirconium silicate, $ZrSiO_4$. Because of its resistance to weathering, and moderate specific gravity (4.68), zircon is a common constituent of a black sands associated with gold placers.

APPENDIX A

PLACER SAMPLING FORMS

PLACER SAMPLE

FIELD RECORD

Sample No. [1] _____ Date _____

State _____ County _____ Serial No. _____

Claim _____

Legal description: Sec. _____ Twp. _____ Range _____

Type of deposit: _____

Type of sample: _____ Dimensions of cut: _____

Place taken: _____

Reference tie: _____

Photograph: _____

INTERVAL   FORMATION _____

___to___ _____

___to___ _____

___to___ _____

___to___ _____

___to___ _____

Bedrock: No ( ) Yes ( ) Type _____

Overburden: Feet _____ Type _____ Water level _____ ft.

Sample Weight: Gross _____ lbs. Net _____ lbs.

Sample Volume: _____ loose measure ( ) bank measure ( )

Boulders: No ( ) Yes ( ) Aver. size _____ Max. size _____ Est. % of bank-run _____

Cement: No ( ) Yes ( ) Remarks _____

Clay: No ( ) Yes ( ) Est. % _____ Remarks _____

Caliche: No ( ) Yes ( ) Est. % _____ Remarks _____

Hard digging: No ( ) Yes ( ) Remarks _____

Sampled by _____ Title _____

Others present _____

Time: Arrive _____ Depart _____ Weather _____

Remarks: _____

_____

_____

Place sketch on back of this sheet.

[1] Carry this number forward to PROCESSING RECORD SHEET, which see.

Sample No. [1] _____

PLACER SAMPLE

PROCESSING RECORD

Sample No. [1] _____   Date _____

Serial No. _____

Dry Weight _____ lbs.; Volume _____; How measured _____

How Processed _____

Processed by _____ Title _____

Others present _____

Visible gold [2] : _____ # 3 colors; _____ # 2 colors; _____ # 1 colors. No visible gold ( );

Gold weight _____ milligrams   Estim. fineness _____

Gold removed: Manually ( ); By amalgamation ( ); Other _____

Character of Gold: Fine or flaky ( ); Coarse or shotty ( ); Smooth ( ); Rough ( );

    Remarks _____

When panning, does Gold tend to ride over top of black sand ( ); or stay down on pan ( ):

    Remarks _____

Does Gold amalgamate readily: Yes ( ); No ( ):

    Remarks _____

Weight of black sand in sample _____; Lbs. black sand per cubic yard _____

Amount of black sand is: Large ( ); Medium ( ); Small ( )

Screen was used in washing process: No ( ); Yes ( ); Size of openings _____

_____ % Oversize; _____ % Undersize:

    Remarks _____

Material washed: Easy ( ); Normal ( ); Difficult ( ):

    Remarks _____

Muddy water indicated by washing: Little ( ); Moderate ( ); Much ( ).

Grade of sluice or rocker _____ in./ft.; Source of wash water _____

Notes: _____

---

[1] Number the same as FIELD RECORD sheet, which see.
[2] #3 colors consist of gold particles weighing less than 1 milligram;
    #2 colors weigh between 1 and 4 milligrams;
    #1 colors weigh over 4 milligrams.
    Weigh and note individual colors weighing 10 milligrams or more.

Sample No. [1] _____

PLACER SAMPLE

DATA SUMMARY SHEET

Date _____

Serial No. _____

State _____ County _____ Legal description _____

Remarks _____

Sheet _____ of _____

Sample Number 1/	Where Taken	Net Weight or Volume	Bank-measure or pan factor	How Processed	Recovered Gold (Mg)	Cents Au @ $ /oz	Cents per Cubic Yard	Notes

Sampled by _____

Processed by _____ Calculated by _____

1/ Sample numbers from Field or Processing Record sheets, which see.

155

FIELD LOG

SHAFT NO. _____ LOCATION _____

Date	Formation		Sampl.	Depth	Cu.Ft. Excavation	Cu.Ft.Box Measure	Gold Recovered	Value Per Cu. Yd.	Workmen		
									Shaft	Hoist	Rocker
		1									
		2									
		3									
		4									
		5									
		6									
		7									
		8									
		9									
		10									
		11									
		12									
		13									
		14									
		15									
		16									
		17									
		18									
		19									
		20									
		21									
		22									
		24									
		25									
		26									
		27									
		28									
		29									
		30									

NOTES:

	DEPTH
	VALUE
	W. L.
	B. R.

156

APPENDIX B

GOLD PRICE AND VALUE DATA

U. S. PRICE OF GOLD

Date of Law	Value Per Fine Oz. Troy
April 2, 1792	$19.393939
June 28, 1834	$20.689656
Jan. 18, 1837	$20.671835
Mar. 4, 1900	$20.671835
Jan. 31, 1934	$35.000000
Mar. 17, 1968	U.S. Treasury regulations published in the Federal Register on March 19,1968, allow domestic gold producers to sell newly mined gold to persons regularly engaged in an industry, profession, or art, who require gold for legitimate, customary, and ordinary use, or persons holding Treasury gold licences. The price of such gold is no longer regulated by the Treasury. The price of monetary gold remains fixed at $35.00 per ounce. The U.S. Mint will no longer purchase gold in the private market nor sell it for industrial, professional or artistic uses. The private holding of gold by U.S. citizens continues to be prohibited except pursuant to existing regulations.

TROY WEIGHT (FOR GOLD & SILVER)

1 pound troy = 12 oz. troy = 5760 grains
$$= 373.2417 \text{ grams}$$

1 ounce troy = 20 pennyweights = 480 grains

1 ounce troy = 31.1035 grams = 31,103 milligrams

1 pennyweight = 24 grains = 1.5551 grams

VALUE OF GOLD, ¢ PER MILLIGRAM

$$= \frac{\$ \text{ Price x fineness x } 0.1}{31,103}$$

Example:

Gold @ $70 per ounce and 825-fine.

$$\frac{70 \times 825 \times 0.1}{31,103} = 0.185¢ \text{ per mg.}$$

MILLIGRAMS OF GOLD EQUIVALENT TO 1¢

$$= \frac{1}{¢ \text{ per mg.}}$$

Using foregoing example:

$$\frac{70 \times 825 \times 0.1}{31,103} = 0.185¢ \text{ per mg.}$$

$$\frac{1}{0.185} = 5.4 \text{ mg. equivalent to } 1¢.$$

GOLD FINENESS AND VALUE EQUIVALENTS (based on $35/oz.)

FINENESS	VALUE PER			
	TROY OZ.	PENNEYWEIGHT	GRAIN	MILLIGRAM
1000	$35.00	$1.75	7.29c	0.112c
975	$34.12	$1.71	7.11c	0.109c
950	$33.25	$1.66	6.93c	0.107c
925	$32.38	$1.62	6.74c	0.104c
900	$31.50	$1.57	6.56c	0.101c
875	$30.63	$1.53	6.38c	0.098c
850	$29.75	$1.49	6.20c	0.096c
825	$28.88	$1.44	6.02c	0.093c
800	$28.00	$1.40	5.83c	0.089c
775	$27.13	$1.36	5.65c	0.087c
750	$26.25	$1.31	5.47c	0.084c
725	$25.38	$1.27	5.29c	0.082c
700	$24.50	$1.22	5.10c	0.079c
675	$23.63	$1.18	4.92c	0.076c
650	$22.75	$1.14	4.74c	0.073c
625	$21.88	$1.09	4.56c	0.070c
600	$21.00	$1.05	4.38c	0.067c

TO FIGURE THE VALUE OF GOLD AT ANY PRICE OR FINENESS:

$$\frac{\text{New price}}{35} \times \frac{\text{Fineness}}{1000} \times 0.112 = \text{¢ per milligram}$$

$$\frac{\text{New price}}{35} \times \frac{\text{Fineness}}{1000} \times 7.29 = \text{¢ per grain}$$

$$\frac{\text{New price}}{35} \times \frac{\text{Fineness}}{1000} \times 1.75 = \text{\$ per penneyweight}$$

APPENDIX C

WATER DATA

WATER REQUIREMENTS

The source, amount, and delivered cost of water are important elements in a placer operation. In many cases they determine the type of equipment or mining method used. Water estimates for new or proposed operations are generally based on experience or working data obtained from comparable operations.

The water required for various working methods varies widely and depends on many factors. Examples that folow are intended only to show the possible range.

a. *Rockers:* A steady flow of 4 or 5 gallons per minute is sufficient to operate a small (1'x4') rocker. Water can be dipped from a barrel where steady flow is not available. Net water consumption may be as low as 50 to 100 gallons per cubic yard, if carefully saved and reused.

b. *Small-scale hand mining:* Where material is loosened by picking, and shoveled into a sluice box by one or two men, 170 to 225 g.p.m. are required for a 12-inch box with steep grade.

c. *Ground sluicing:* Water duty varies widely but may range between 1/10 and 3/4 cubic yard per miners' inch-day at small mines. This would be equivalent to about 22,000 to 162,000 gallons per cubic yard.

d. *Hydraulicking:* Water duty varies widely and reflects the coarseness of gravel, degree of cementing, height, of bank, grade of bedrock, available head, etc., and is commonly between 1/2 and 7 cubic yards per miner's inch-day. This would be equivalent to about 2,000 to 32,000 gallons per cubic yard. The better efficiencies are obtained at large, well-equipped mines. Small, 1 or 2-monitor mines operated by individual owners or lessees, usually have a water duty of less than 1 cubic yard per miner's inch-day.

e. *Stationary washing plants:* These are typically owner-operated plants, fed by a dragline or a small power shovel. Most employ a trommel or other screening device ahead of the sluice. Incomplete figures indicate a range of 650 to 2,000 gallons per cubic yard.

f. *Moveable washing plants and dryland dredges:* In same category as stationary plants and same remarks apply. Water requirements ranging from 480 to 3200 gallons per cubic yard have been noted. Plants equipped with Ainlay bowls (in place of sluices) generally have good water economy.

g. *Dragline dredges:* Net water required for washing gravel and maintaining the pond is governed by the amount of clay, porosity of the gravel, and other factors. Wash water which is commonly supplied by an 8-inch centrifugal pump working against a 40-foot pressure head, may range between 570 to 2,500 gallons per cubic yard.

h. *Bucket-line dredges:* Water in circulation aboard a dredge may range from 3,500 g.p.m. to over 10,000 g.p.m. depending on digging capacity of dredge and type of material being washed. Dredges are usually provided with independent high pressure and low pressure water systems, the high pressure being used for screen sprays and bucket nozzles, and the low pressure for the gold-saving tables and general service. When working in land-locked ponds, a fresh water input of 1,000 g.p.m. to more than 2,000 g.p.m. will be needed to replace muddy water which must be pumped out of the pond (to prevent excessive mud build-up) and to maintain pond level.

UNITS OF WATER MEASUREMENT

1 gallon (gal.) = 231 cubic inches .
          = 0.1337 cubic feet.
1 gallon of water weighs 8.33 pounds.
1 million gallons (m.g.) = 3.0689 acre feet.
1 cubic foot (cu. ft.) = 1728 cubic inches.
          = 7.48 gallons.
1 cubic foot of water weighs 62.4 pounds.
1 acre foot (ac. ft.) = amount of water required to cover one acre one foot deep.
          = 43,560 cubic feet.
          = 325,850 gallons.
          = 12 acre inches.
1 gallon per minute (g.p.m.) = 0.00223 cubic feet per second.
          = 1440 gallons per day (24 hrs.).
1 million gallons
per 24 hours (m.g.d.) = 1.547 cubic feet per second.
          = 695 gallons per minute.

1 cubic foot

  per second (sec. ft.) = 7.48 gallons per second.

                          = 448.8 gallons per minute.

                          = 646,272 gallons per day (24 hrs.).

                          = .992 acre inch per hour.

                          = 1.983 acre feet per day (24 hrs.).

                          = 40 miner's inches (legal value) in Arizona, California, Montana and Oregon.

                          = 50 miner's inches (legal value) in Idaho, Kansas, Nebraska, Nevada, New Mexico, North Dakota, South Dakota and Utah.

                          = 38.4 miner's inches in Colorado.

                          = 35.7 miner's inches in British Columbia.

1 miner's inch (mi. in.) = 11.25 gallons per minute when equivalent to 1/40 second foot.

                          = 9 gallons per minute when equivalent to 1/50 second foot.

1 miner's inch-day (24 hrs.) = 16,200 gallons when equivalent to 1/40 second foot.

1 miner's inch-day (24 hrs.) = 12,960 gallons when equivalent to 1/50 second foot.

APPENDIX D
PLACER DRILLING DATA

Setting Up The Drill

THE drill is placed in the desired location for the hole, leveled, and the derrick raised. One end of the sand line is attached to the vacuum sand pump; the other end is placed over the sand line sheave in the derrick and attached to the sand line drum. The drill line is placed over the crown sheave of the derrick, laced through the walking beam, and attached to the mainline drum, the cable socket being on the other end of the line.

For most placer operations, a string of tools usually consists of a rope socket, a stem, and a bit. When assembling a string of tools, lay the stem on the ground at the front of the drill; screw on the rope socket with the attached cable; then screw the bit on the lower end of the stem. A light wire brush and gasoline is recommended to clean the threads before assembling. A few drops of light oil are desirable on the threads.

Great care must be exercised in assembling a string of tools to see that the joints are tight. Screw the tool joints up by hand, then apply tool wrenches and set the joint up firmly by means of the chain wrench bar. Remember, *most fishing jobs are due to improper setting of joints.*

Many drillers use a stone in polishing the shoulder portion of the box and pin, before assembling. After a joint has been set up perfectly solid, a mark is sometimes made with a sharp cold chisel, half on the pin collar and half on the box. Each successive time this joint is screwed up, the mark on the box should go a little farther past the mark on the pin collar. By this method one can check and see that the joint is set up firmly. If, at the next fitting the joint does not go far enough, it is indicative that there is dirt on the face of the joint – or in the threads, which dirt must be removed.

After the stem and bit have been assembled, the tools are hoisted, the operator being careful not to kink the cable where it enters the socket. Check the frame to see that the tools hang directly above the center line of the drill and approximately 24 inches out from the front of the drill. You are now ready to start the hole.

Starting a Hole

AT THE SITE marked for drilling, a hole is dug approximately 16 inches deep. The gravel removed is examined to see whether it contains gold. A drive shoe is placed on one end of a length of drive pipe and a drive head on the other.

After carefully measuring the length of the pipe and shoe they are placed in the hole in a vertical position with the drive shoe on the bottom. Dirt is packed around the pipe to hold it in position. Where long lengths of casing are used, a well is usually dug directly in front of the drill. This is very difficult to do in wet ground.

DRIVING THE PIPE

THE TOOLS are lowered, allowing the bit to enter the casing, the drive clamps are clamped on the square on the bit. It is important to see that the drive clamp bolts are kept tight. Loose nuts mean certain breaking of bolts.

The engine of the drill is started and operated at slow speed.

The drive clamps are lowered to within three inches of the drive head, the spudding lever is moved to the "On" position, and the casing gently tapped into the ground. The impact blow and feed are adjusted by gently raising the brake lever.

If the top soil is valueless, the pipe is driven to gravel. When the drive head reaches a point a few inches from the ground, it is removed and another length of pipe is carefully measured and coupled to the first length, care being taken to couple the pipe so that they butt in the center of the coupling. The drive head is then placed on the top of the next section of pipe and the operation resumed, core in the meantime is being removed as the pipe is driven deeper.

Whenever pipe is added, the threads are carefully cleaned and then greased with graphite and linseed oil and securely tightened by means of chain tongs. It is very important to grease the pipe and not the couplings. The portion of the thread one-quarter inch back from the end of the pipe is the part usually lubricated.

As soon as the hole has reached a depth sufficient for the bit and stem to enter the casing, the drive clamps are placed on the top square of the stem. After driving pipe to the desired depth, the drive clamps are removed.

Records are made of the depth of the drive pipe and also the core. By starting at the bit and measuring back along the stem, the total length of pipe is marked with chalk or a piece of string on the line. The stem is lowered into the pipe and the distance between chalk mark and the top of the pipe represents the core remaining in the hole.

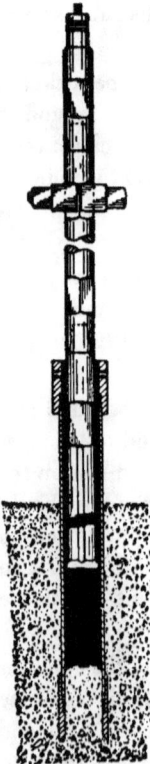

DRIVING

DRILLING OPERATIONS

DRILLING PROCEDURE

DRILLING proceeds in the following manner:

The driller measures and reports the length of core to the panner. This, and the depth of casing in the ground, are recorded in the log book.

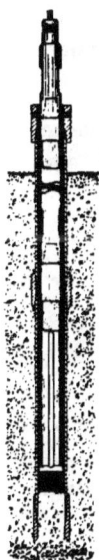

Water is poured into the casing, the spudding lever thrown in the "On" position, and the core chopped up.

Some operators place a spoonful of lye in the drill hole when churning. Lye cuts any grease from the fine gold and prevents it floating away. There are times when there is a certain amount of vegetable oil in the ground or grease from pipe threads or wire line.

In placer testing it is always customary to try to drive the casing ahead of drilling. Three or four inches of core are always left in the casing to form a plug. This is contrasted with other types of drilling where the churning is done below the casing.

When large boulders are encountered, it is necessary to drill below the drive shoe; then the operator should check to see that the water level in the pipe is at least as high as the water plane in the ground, to prevent any values being carried into the casing.

CHURNING

When drilling through boulders greater speed can be obtained by using an ordinary rock bit or a four-wing type bit, than with a placer bit.

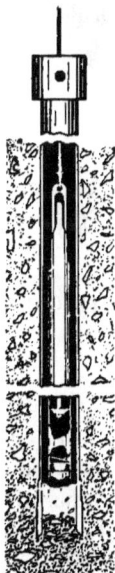

PUMPING OUT THE CORE

THE TOOLS are hoisted out of the pipe and held out of line of the drill casing by means of the tool guide. While the tools are being hoistered, a bucket-of water is usually poured on the rope and stem to wash off any values that might be clinging to the tools. The sand pump is raised by pulling on the sand pump lever. When the lever is released, a brake keeps the sand drum from turning, and holds the pump in position. When the sand pump lever is half way between the brake position and the hoist position, the drum is free, allowing the pump to fall freely to the bottom of the hole.

When the pump is at the bottom of the hole, the piston type plunger is at the bottom of the pump. By pulling on the sand drum lever, the plunger is rapidly raised, creating a strong vacuum, sucking in the gold, sand, mud, cuttings, etc. The efficiency of a sand pump depends on the speed with which

PUMPING

the plunger is raised.

The foot valve in the bottom of the pump prevents material from escaping. Holding the sand drum lever out and keeping the sand drum engaged, brings the pump containing the core to the surface. Engineers very often drop several lead shot in the hole without the operator's knowledge, and then count them after the pumping to see that everything was removed.

It is common practice to pump before and after driving. In loose ground it is often possible to pump out the core immediately after driving. When the two pumpings are made before and after driving, they are caught in the same bucket and concentrated in one operation. The pumpings are ordinarily emptied into a mud box.

When prospecting a deposit having a deep overburden of valueless material, a one-inch rod is driven into the ground and as the pump is brought up, the foot valve of the sand pump is set on the rod, thus opening the valve and washing the contents out on the ground. The driller must be careful, however, not to throw away the contents when gravel is reached. Extreme care should be exercised so that the last three or four inches of the core is left in the bottom of the drive pipe to form a plug.

After the stem has been lowered into the pipe, and the depth of the remaining core checked and recorded in the log book, the tools are raised. The drive clamps are bolted on and the operations repeated. After the first drive, the clamps are bolted on the top square. It is impossible to do this at the start, as bit would strike the core when driving.

PIPE PULLING

A SPECIAL FEATURE of the Hillman Drills is their remarkable pipe pulling ability.

Using a top puller, the operation of pulling the pipe is as follows: The drive head is removed from the top of the casing. The knocking head is slipped over the pipe pulling jar. The rope socket is attached to the pin of the pipe pulling jar. Some operators prefer to use the rope socket and stem above the pipe pulling jar but a sharper blow seems to be obtained when it is below. The knocking head is screwed on the top of the pipe in place of the driving head. The drill is placed in the spudding position and the string of tools is raised far enough so that by means of the spudding action they are thrown at the driving head. A sharp blow is obtained rather than a straight pull. This is exactly the reverse action of pipe driving, the blow being directed at the lower face of the knocking head instead of the top of the driving head.

PULLING RING

PULLING

Sometimes considerable time is saved by using a hand hoist in conjunction with the bumping action. By thus having a tension on the line, the pipe does not have a chance to drop back after each blow and a continuous upward tension on the pipe is maintained.

In extreme instances where the burden of jarring the pipe is too great for normal blow delivered and the upward tension of a hand hoist line is insufficient, a pipe pulling ring should be used. By placing the ring on the casing-head and hand jacks beneath the flange on the ring, a tremendous upward pressure is added to the jarring effect of the stem which starts the pipe upward.

PIPE PULLING WITH CASING SPEAR

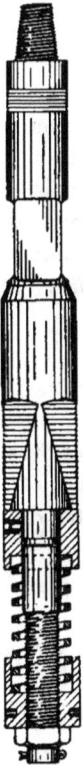

THE C. KIRK HILLMAN Company makes a three-jaw casing spear. Three jaws are used to eliminate distortion of the pipe as is usually the case when two jaws are used. Four jaws do not work as well as three jaws for six-inch casing and smaller, for not enough stock is then left in the center of the spear to withstand the constant hammering.

The operation of pulling the pipe by means of the spear is as follows: The string of tools is assembled, consisting of a rope socket, a stem, a set of long stroke jars and the trip casing spear. The jaws of the spear are set by using the spider furnished with the spear for pulling back the jaws and compressing the spring. The anvil block directly on top of the spring has a key which slides into a notch on the lower end of the spear; this key is pressed in. The spider is then removed. The string of tools is then lowered into the casing. Pulling is usually done from the last length. When the spear is in the last length of pipe it is hoisted up, sinking the jaws into the casing. The drill is placed in the spudding position and the string of casing jarred upwards. To release the casing, the spear is struck a sharp blow from the top. This causes the anvil block to strike the top of the jaws, which releases the spear. The tools are then hoisted out of the casing. The lengths of casing out of the ground are uncoupled and the operation repeated.

SPEAR

JARS

TREATMENT OF THE CORE

WHEN the sand pump is hoisted out of the casing, the helper, or panner, grasps the lower end of the pump, walks back from the drill, and lays the lower end on the saddle of the dump (or mud) box. The bail end of the sand pump is then lowered into the bottom of the dump (or mud) box. When the pump in this position the foot valve is approximately on an even height with the panner's eye. While the pump is in this jackknifed position, it is thoroughly washed both inside and out. A dipper, formed by fastening a handle on a gallon tin can, is used for pouring water in the foot valve and slushing out the material. The pump is hoisted and jackknifed in the other position. The lower end of the pump on the previous washing is now the top. The second washing is not always essential, but is advisable, especially when working in rich ground. Directly below the lower end of the mud box is located a tub containing the volume bucket.

MUD BOX OR DUMP BOX

THE LEGS and sides of the dump box are made of surfaced fir, two inches by four inches. The trough is formed of 16-gauge steel. The top end of the box is welded solid. The lower end is also welded and fitted with an adjustable gate which can be set to control the flow into the volume bucket and avoid splashing.

THE VOLUME BUCKET

THE Volume Bucket, holding one cubic foot, is 13½ inches in diameter and 12 inches in height. The measuring stick is calibrated in tenths and hundredths of a cubic foot. A good practice is to measure the total quantity of material pumped from each hold and to compare the results from all holes drilled in similar material. A fair average can be obtained and used as a check on each separate property. Some operators calculate twenty cubic feet of material so measured to represent one cubic yard actually drilled.

The volume bucket measurement entails complete disintegration of the material and its reaggregation before measurement takes place. The escaping slimes may or may not fill the interstices of the sand with doubtful increase of the volume of the sand. Therefore, measurements made by this method will not *always* check with the theoretical volume or the volume as measured by the rise of the core in the pipe.

The material in the volume bucket is thoroughly stirred to break up any sticky lumps and then the water is poured off. Records are made in the log book of the measured volume of the core Directly beside the volume bucket is the panning table.

THE PANNING TABLE

THE Panning Table is usually 28 inches high by 30 inches wide by 60 inches long. It has two wash tubs on top. A safety pan is placed in the bottom of each tub. A shelf is directly under the table top to hold the dish for holding the concentrates. The grizzly pan and the regular pan are also part of the equipment for the panning table. The grizzly pan has three-eight-inch holes. This allows the small material to wash through, and speedily eliminates the coarse material.

The material is taken out of the volume bucket with the gold scoop and placed in the grizzly pan which is in the second pan. The pans are usually filled two-thirds full. The stones and large cuttings are washed clean while the small particles fall through the grizzly pan into the second pan. The rocks are looked over and examined for nuggets before thrown out and the rough panning is done in the first tub. The fine panning is done in tub No. 2, where the water is clear. By having the water slightly warmed, the working conditions are more pleasant for the operator and also the clay tends to disintegrate more readily. The panner classifies the gold in one, two, and three colors. No. 3 is the finest, and consists of all particles weighing less than 1 Mg; No. 2 is gold consisting of all particles weighing between 1 Mg. and 4 Mg.; while No. 1 gold is any particle weighing over 4 Mg. The colors are recorded in the log book. The concentrate is placed in a bottle, and marked with a Line Number, Hole Number, and Depth.

THE SAMPLE BOTTLES

FOR handling samples, some of the companies use small bottles similar to a vaseline jar. The cover is made of aluminum and of the screw type. The top of the cover is flat and has a dull finish, similar to a sand blasted surface. Writing with lead pencil is very legible on this type surface. The Line Number, Hole Number, Depth, and the panner's signature are all written on top of the bottle.

After all values have been recorded and the bottles are again sent out in the field, the writing is removed by a scouring action with the thumb, using sand and water. This method eliminates the danger of lost labels. Remember: reliable evidence inspires confidence in the crews' reports.

THE ROCKER

DIRECTLY beside the panning table is the rocker. The rocker, as used for checking drill samples, is very similar to the one used by the small individual placer operator. Some of the differences are: three-eight-inch holes in place of half-inch-holes in the punched plate; punched portion of plate only two-thirds of half size used in the ordinary rockers, preventing all the material from being washed through with the first dipper of water. As the rocker is cleaned up so often, only two or three riffles are necessary.

TREATMENT OF GOLD

A FEW of the operators take the bottles of concentrate from the hole, carefully pan them down as far as possible, dry them, remove the magnetic material with a magnet, blow the remaining black sand off and then weigh the gold.

The usual procedure is to use a six-inch gold pan, pan the material down to a high concentrate, and them amalgamate the gold. This is done by rubbing the black sand and concentrate with mercury, then carefully panning off the black sand. This is usually caught in another pan and checked to see that no values have been lost. The samples are then usually placed in bottles marked for future reference. At this time it is well to qualitatively inspect the contents of the pan and black sands for platinum or unusual amounts of any of the other valuable metals or minerals which are apt to be found in placer deposits.

There are times when the gold does not readily amalgamate, due to oil or grease from the sand pump, pipe joints, vegetation, et cetera. Adding a little caustic potash usually eliminates this trouble. To amalgamate the gold, some operators add one or two drops of dilute nitric acid to the amalgam and concentrates and stir vigorously was a glass stirring rod until all the gold has been absorbed by the mercury.

The ball of amalgam is then carefully placed in a porcelain annealing cup or test tube. Dilute nitric acid (specific gravity 1.42, plus an equal volume of water) is added to the cup and the contents heated over the flame of an alcohol lamp. The mercury goes into solution with the nitric acid, leaving the natural alloy of gold and silver. The acid should be heated only enough to accelerate the reaction. Excessive or continual heating will dissolve the alloy in the gold and change its fineness. This loss of weight may cause an error, as the values are computed on the basis of natural fineness of the gold. The mercury nitrate is carefully washed from the gold and silver alloy, using a wash bottle with hot water. After adding a drop or two of alcohol to the annealing cup to prevent sputtering, it is heated over the alcohol lamp to a red heat or to a point as hot as the lamp will allow. The operation removes the moisture and burns away any carbon or lint which would salt the weight of the gold. The gold is then weighed and values recorded in the log book.

LOG BOOK

An accurate systematic record should be kept of each hole drilled. Neat, accurate and carefully kept, and signed drillers logs are of the utmost importance and should be as carefully prepared as banking or legal papers. The reliability of a report on any placer deposit depends on the care with which the examination is conducted and recorded.

The driller and panner have a good opportunity to examine all the conditions relative to the deposit. All factors which may in any way have any bearing on the cost of the operation should be noted in the logs. Any corrections or compensation for lack of core should also be made in the field at the time, as later in the office, engineers might not fully understand all the conditions.

Explanations of Headings on Log Sheet

The following data is usually recorded on the log sheets:
1—Name of property
2—Location
3—Date
4—Line Number
5—Hole Number
6—Elevation

7—Time: Entries are made in this column of the time of day each pumping is made. At the bottom of the page the time consumed in drilling, pulling, moving, repairs, and delays is recorded. These records are helpful in determing the cost of prospecting similar ground.

8—Depth of the cutting edge of the drive shoe: The last item in this column is the total depth drilled.

9—Depth of pumping: Care must be taken to see that pumping is not continued farther than within two inches of the lower edge of the drive shoe until bedrock has been reached or when a boulder is encountered. Notation should always be made when drilling below the shoe.

10—Entries are made of each individual drive.

11—The rise of the core in the pipe for each drive: This is the distance from the bottom of the drive shoe to the top of the core.

12—Core after pumping: The amount of core left in the pipe is naturally less than before pumping. This is the distance from the bottom of the drive shoe to the top of the core and is termed the plug. It usually varies from two to four inches.

13—The length of core removed: The difference between the core before pumping and the core after pumping shows the rise of the core in the pipe. Some operators use this in computing the volume. Volume computed by this method is known as the core volume by pipe measurement.

14—Volume Bucket measurement. This is exactly as the name implies, the volume as measured in the volume bucket.

15—Classification of colors: The number of colors are classified in the No. 1, No. 2, and No. 3 sizes. No. 3 is the finest and consists of all particles weighing less than one milligram. No. 2 is the gold consisting of all particles weighing between one milligram and four milligrams. No. 1 gold is any particle weighing over four milligrams. These are recorded on the lines opposite the depths of drive, so that the various pay streaks will be known.

16—Estimated weight of gold: The panner becomes very proficient in estimating weight. He records the weight of the different colors in the log on lines corresponding with the depths at which they are found. This is helpful in following pay streasks.

17—Formation: The nature of the formation corresponding with the various depths is also noted. Very often the pay streak has a distinctive color. It is advisable to make a notation of this. The size and quantity of boulders have a definite bearing on how the property should be worked. Properties containing an unusual amount of clay require special washing equipment to prevent the clay from going through the boxes and picking up the gold. Some properties contain black sand which has a tendency to pack the riffles and decrease the amount of material which can be handled.

19—Depth and nature of over-burden: Very often the over-burden is removed by a different mehtod than that used for working of the gravel. Information such as buried logs, large boulders, etc., have a direct bearing on the cost, and records should be made of them.

20–Labor conditions, transportation facilities.

21–Depth of the pay gravel.

22–Depth to bed rock.

23–Nature of bed rock.

24–Depth of pay in bed rock.

25–Diameter of the drive shoe.

26–Theoretical volume of the core removed.

27–Measured volume.

28–Weight of gold in milligrams. (mg.)

29–Fineness of the gold.

30–Constant used in making calculations.

31–Value in cents per cubic yard.

32–Price of gold at which values were computed.

33–Signatures of the driller, panner and helper.

A list of abbreviations is included at the top of the log sheet which simplifies the recording.

DRILLERS FIELD LOG

SHEET No.................. SHEETS THIS HOLE..................

LINE No.................. HOLE No..................

DATE START.................. DATE FINISH..................

TIME OF PUMPING	DEPTH OF								CORE RISE Inches	PLUG AFTER PUMPING Inches	DRILLED AHEAD Inches	CORE FOR DRIVE	VOLUME Measured by VOLUME BUCKET	Number of COLORS			ESTIMATED WEIGHT	COLOR OF SLUDGE	FORMATION
	PIPE		PUMPING		DRIVE									1	2	3			
	Feet	Inches	Feet	Inches	Feet	Inches													

DRILLING REPORT

Name of Property

Location of Property

Line No   Hole No.

Location of Hole

Depth of Tails

Depth of Overburden

Depth of Gravel

Depth of Bedrock

Depth of Dredge Section

Depth of Hole: Casing   Open

Gold @   per Ounce   per mg.

Black Sands: Much   Some   Little

Coarse   Medium   Fine

Size Casing

Size Shoe

Volume from Shoe Constant

Volume from Core Rise

Volume from Bucket

Volume from Water Displacement

Volume for Estimate

Gold, Actual Wt.   Mgs.

VALUE: CENTS PER CU. YD.

CROSS SECTION OF MATERIAL

ELEVATIONS

Surface

Bedrock

Water Level

From	To	Mgs., Constant	, Cu. Yds.	, ¢ per Yd.
From	To	Mgs., Constant	, Cu. Yds.	, ¢ per Yd.
From	To	Mgs., Constant	, Cu. Yds.	, ¢ per Yd.
From	To	Mgs., Constant	, Cu. Yds.	, ¢ per Yd.
From	To	Mgs., Constant	, Cu. Yds.	, ¢ per Yd.
From	To	Mgs., Constant	, Cu. Yds.	, ¢ per Yd.
From	To	Mgs., Constant	, Cu. Yds.	, ¢ per Yd.
From	To	Mgs., Constant	, Cu. Yds.	, ¢ per Yd.

TIME LOG

Drilling	Hrs.	Driller	
Pulling	Hrs.	Panner	
Moving	Hrs.	Helper	
Repairs	Hrs.	Field Supt.	
Delays	Hrs.	Metallurgist or Engineer	
Total Time	Hrs.	Checked by:	

SUMMARY SHEET

Line _____   Hole _____

SAMPLE NUMBER	POSITION	DEPTH FEET	VOLUME CUBIC YARDS	WEIGHT AU. IN MGS.	VALUE AU. AT $	VALUE CENTS PER YARD	DEPTH TIMES VALUE	CUMULATIVE VALUE
TOTAL								

REMARKS

183

TABLE OF MULTIPLYING FACTORS FOR DETERMINING GOLD-VALUE PER CUBIC YARD FOR VARIOUS DIAMETER DRIVE SHOES

To determine the value in cents per cubic yard, multiply the number of Mgs. of gold recovered from the hole by the factor shown in the following table which corresponds to the depth of hole and the proper diameter of drive shoe. These are computed from the following formula:

$$\text{Value per cu. yd.} = \frac{\text{Mg. recovered} \times \text{price per Mg.}}{\text{Volume (in cubic yards)}}$$

This value is equal to K x Mg. recovered, where K is the proper multiplying factor. The factors are based on the following values:

VALUE OF GOLD $35.00 PER OUNCE. FINENESS OF GOLD 1000

DEPTH OF DRILL HOLES EXPRESSED IN FEET — DIAMETER OF DRIVE SHOE EXPRESSED IN INCHES

These factors are computed with the assumption that the volume of core obtained is represented by a cylinder whose length is equal to the depth of hole and the diameter of the drive shoe. The volume of ground obtained is usually somewhat less than the volume represented by this cylinder. When this is true, the gold actually recovered is credited to a larger volume of ground than that from which it was actually obtained. The actual value per cubic yard is then slightly greater than that shown by calculating using the following table.

Depth to Bedrock	5¼" Shoe	6½" Shoe	7½" Shoe	9¾" Shoe
6	3.368	2.1972	1.651	.9766
6.5	3.109	2.0282	1.523	.9015
7	2.888	1.8833	1.415	.8371
7.5	2.694	1.7577	1.320	.7813
8	2.526	1.6479	1.237	.7324
8.5	2.378	1.5509	1.165	.6894
9	2.245	1.4648	1.100	.6510
9.5	2.127	1.3877	1.042	.6168
10	2.021	1.3183	.9903	.5860
10.5	1.925	1.2555	.9430	.5581
11	1.837	1.1985	.9003	.5327
11.5	1.757	1.1463	.8612	.5096
12	1.684	1.0986	.8250	.4883
12.5	1.619	1.0547	.7922	.4684
13	1.554	1.0141	.7617	.4507
13.5	1.498	.9767	.7335	.4340
14	1.444	.9416	.7074	.4186
14.5	1.393	.9092	.6829	.4042
15	1.347	.8789	.6602	.3906
15.5	1.304	.8505	.6389	.3781
16	1.263	.8239	.6187	.3663
16.5	1.225	.7990	.5999	.3551
17	1.189	.7755	.5824	.3447
17.5	1.155	.7533	.5658	.3348

Depth to Bedrock	5¼" Shoe	6½" Shoe	7½" Shoe	9¾" Shoe
33.5	.6032	.3935	.2955	.1749
34	.5943	.3877	.2912	.1723
34.5	.5857	.3821	.2870	.1698
35	.5774	.3767	.2828	.1674
35.5	.5692	.3714	.2789	.1651
36	.5613	.3662	.2750	.1628
36.5	.5536	.3612	.2713	.1605
37	.5462	.3563	.2675	.1584
37.5	.5389	.3516	.2640	.1563
38	.5318	.3469	.2604	.1542
38.5	.5249	.3424	.2571	.1522
39	.5182	.3380	.2539	.1503
39.5	.5116	.3337	.2506	.1483
40	.5052	.3296	.2475	.1465
40.5	.4990	.3255	.2445	.1447
41	.4929	.3215	.2415	.1429
41.5	.4869	.3177	.2385	.1412
42	.4811	.3139	.2358	.1395
42.5	.4755	.3102	.2329	.1379
43	.4699	.3066	.2302	.1363
43.5	.4645	.3031	.2275	.1347
44	.4593	.2996	.2250	.1332
44.5	.4541	.2963	.2224	.1317
45	.4491	.2930	.2200	.1302

45.5	.4441	.2897	.2176	.1288
46	.4393	.2866	.2152	.1274
46.5	.4346	.2835	.2129	.1260
47	.4300	.2805	.2106	.1247
47.5	.4254	.2775	.2084	.1234
48	.4210	.2746	.2063	.1221
48.5	.4167	.2718	.2041	.1208
49	.4124	.2690	.2020	.1196
49.5	.4082	.2663	.2000	.1184
50	.4042	.2637	.1980	.1172
50.5	.4002	.2611	.1961	.1160
51	.3962	.2585	.1941	.1149
51.5	.3924	.2560	.1922	.1138
52	.3886	.2535	.1903	.1127
52.5	.3849	.2511	.1886	.1116
53	.3813	.2487	.1868	.1106
53.5	.3777	.2464	.1850	.1095
54	.3742	.2441	.1833	.1085
54.5	.3708	.2419	.1816	.1075
55	.3674	.2397	.1800	.1065
55.5	.3641	.2375	.1784	.1056
56	.3609	.2354	.1768	.1046
56.5	.3577	.2333	.1752	.1037
57	.3545	.2313	.1737	.1028
57.5	.3514	.2293	.1722	.1019
58	.3484	.2273	.1707	.1010
58.5	.3454	.2254	.1692	.1002
59	.3425	.2234	.1678	.09932
59.5	.3396	.2216	.1664	.09848
60	.3368	.2197	.1651	.09766
60.5	.3340	.2179	.1638	.09636

18	1.123	.7324	.5500	.3255
18.5	1.092	.7126	.5351	.3167
19	1.064	.6938	.5215	.3084
19.5	1.036	.6761	.5077	.3005
20	1.011	.6592	.4950	.2930
20.5	.9858	.6431	.4829	.2858
21	.9621	.6278	.4717	.2790
21.5	.9399	.6132	.4605	.2725
22	.9185	.5992	.4500	.2664
22.5	.8981	.5859	.4400	.2604
23	.8786	.5732	.4304	.2548
23.5	.8599	.5610	.4216	.2495
24	.8420	.5493	.4125	.2442
24.5	.8248	.5381	.4041	.2392
25	.8083	.5273	.3960	.2344
25.5	.7925	.5170	.3882	.2298
26	.7772	.5071	.3810	.2254
26.5	.7626	.4975	.3740	.2211
27	.7485	.4883	.3670	.2170
27.5	.7348	.4794	.3600	.2131
28	.7217	.4708	.3540	.2093
28.5	.7091	.4626	.3475	.2056
29	.6968	.4546	.3417	.2021
29.5	.6850	.4469	.3358	.1986
30	.6736	.4394	.3300	.1953
30.5	.6626	.4322	.3246	.1921
31	.6519	.4253	.3194	.1890
31.5	.6415	.4185	.3142	.1860
32	.6315	.4120	.3094	.1831
32.5	.6218	.4056	.3047	.1803
33	.6124	.3995	.3000	.1776

USE OF HILLMAN MULTIPYING FACTORS
FOR DETERMINING GOLD VALUE PER CUBIC YARD

Using the theoretical value: Suppose that from a 40-foot hole using a 7½" drive shoe, 100 milligrams of gold are recovered. From the table of multiplying factors on the preceding page, find opposite the depth of 40 feet the factor .2475. Also suppose the relative fineness of the gold is 900. The value of the hole would be found as follows:

$$100 \text{ x } .2475 \text{ x } \frac{900}{1000} = 22.3\text{c per cubic yard.}$$

The above value would be correct providing a 100% core had been obtained.

Using a correction factor: The theoretical core rise per foot of drive will be found in the table of prospect drilling factors (see following page). In the above example it will be found that 40 feet x 20.3" = 812", the theoretical core rise. But suppose the actual measured core rise was 893 inches. The value in cents per cubic yard based on a 100% core would then be adjusted by multiplying it by $\frac{812}{893}$, the ratio of actual to theoretical core rise. In our example the adjusted value would be

$$22.3\text{c x } \frac{812}{893} = 20.3\text{c per cubic yard.}$$

The same type of adjustment can be made using volume measurements in place of core rise. Some engineers calculate the adjustment both ways and apply the one giving the least plus correction. There are times when the correction is applied only to certain portions of a drill hole, that is, to sections containing significant values.

PROSPECT DRILLING FACTORS

	4" Drive Pipe Inside dia. 3.826"		5¼" Drive Shoe Cutting Edge dia. 5¼"	
Cross-sectional Area	11.4969	Sq. In.	21.6475	Sq. In.
Volume in Cu. In. per Ft. depth	137.9628	Cu. In.	259.7700	Cu. In.
Volume in Cu. Ft. per Ft. depth	.079835	Cu. Ft.	.1503299	Cu. Ft.
Volume in Cu. Yds. per Ft. depth	.00295695	Cu. Yds.	.00556777	Cu. Yds.

Depth to which 5¼" drive shoe must be driven to cut out a theoretical volume of one cubic yard .... 179.6'

Theoretical rise in 4" drive pipe using a 5¼" drive shoe driven to a depth of one foot .... 22.6"

	5" Drive Pipe Inside dia. 4.813"		6½" Drive Shoe Cutting Edge dia. 6½"	
Cross-sectional Area	18.1938	Sq. In.	33.1832	Sq. In.
Volume in Cu. In. per Ft. depth	218.3255	Cu. In.	398.1978	Cu. In.
Volume in Cu. Ft. per Ft. depth	.1263394	Cu. Ft.	.2304585	Cu. Ft.
Volume in Cu. Yds. per Ft. depth	.0046795	Cu. Yds.	.0085355	Cu. Yds.

Depth to which 6½" drive shoe must be driven to cut out a theoretical volume of one cubic yard .... 117.1'

Theoretical rise in 5" drive pipe using a 6½" drive shoe driven to a depth of one foot .... 21.9"

PROSPECT DRILLING FACTORS

	6" Drive Pipe Inside dia. 5.761"		7½" Drive Shoe Cutting Edge dia. 7½"	
Cross-sectional Area	26.0666	Sq. In.	44.1788	Sq. In.
Volume in Cu. In. per Ft. depth	312.7997	Cu. In.	530.1450	Cu. In.
Volume in Cu. Ft. per Ft. depth	.181019	Cu. Ft.	.3067968	Cu. Ft.
Volume in Cu. Yds. per Ft. depth	.00670433	Cu. Yds.	.0113628	Cu. Yds.

Depth to which 7½" drive shoe must be driven to cut out a theoretical volume of one cubic yard .... 88.00'

Theoretical rise in 6" drive pipe using a 7½" drive shoe driven to a depth of one foot .... 20.3"

	8" Drive Pipe Inside dia. 7.625"		9¾" Drive Shoe Cutting Edge dia. 9¾"	
Cross-sectional Area	45.6636	Sq. In.	74.6621	Sq. In.
Volume in Cu. In. per Ft. depth	547.9632	Cu. In.	895.9450	Cu. In.
Volume in Cu. Ft. per Ft. depth	.3171086	Cu. Ft.	.5184867	Cu. Ft.
Volume in Cu. Yds. per Ft. depth	.0117447	Cu. Yds.	.01920321	Cu. Yds.

Depth to which 9¾" drive shoe must be driven to cut out a theoretical volume of one cubic yard .... 52.08'

Theoretical rise in 8" drive pipe using a 9¾" drive shoe driven to a depth of one foot .... 19.6"

APPENDIX E

GENERAL INFORMATION AND COST DATA

APPROXIMATE WEIGHTS OF EARTH AND GRAVEL

ORDINARY GRAVEL – bank measure	3,000 lbs. per cubic yard
DRY GRAVEL – loose measure	2,000 lbs. per cubic yard
DRY SAND – loose measure	2,200 lbs. per cubic yard
ORDINARY EARTH – bank measure	3,000 lbs. per cubic yard
DRY EARTH – loose measure	2,000 lbs. per cubic yard
CLAY – bank measure	3,200 lbs. per cubic yard

APPROXIMATE SWELL OF EARTH AND GRAVEL[1]

ORDINARY GRAVEL	20 to 30%
CEMENTED GRAVEL	40%
SAND AND GRAVEL	15%
GRAVEL AND CLAY	35%
LOAM	20%
DENSE CLAY	50%

[1] The volume increases shown here may change appreciably under varying conditions, the amount of swell being influenced by the moisture content, size of gravel and its ratio to sand, and other conditions.

GRAIN SIZE

NAME OF PARTICLES	AVERAGE DIAMETER IN MILLIMETERS[1]
Boulders	Greater than 256 mm (10")
Cobbles	64 mm to 256 mm (2½" to 10")
Pebbles	4 mm to 64 mm (3/16" to 2½")
Gravel	Greater than 2 mm
Sand	2 mm to 1/16 mm
Silt	1/16 mm to 1/256 mm
Clay	Less than 1/256 mm

[1] 25.4 millimeters (mm) = 1 inch.

CHARACTER OF PLACER GOLD
related to
DISTANCE FROM SOURCE

It is generally recognized that the shape and angularity of placer gold may serve as a measure of distance traveled. The heavier and more hackly, the nearer the source. The more water-worn and scaly, the further it has traveled from its source. While this can be accepted as fact, the near-absence of statistical data precludes the formulation of specific rules. The following figures should be used with this in mind.

Distance from Source	Nature of Gold [1]
5 miles	Rough nuggety
8 miles	Small nuggety, water-worn
11 miles	Fine granular
25 miles	Fine scaly

[1] Gold in Adelong Creek, New South Wales. Data from article by N. H. Fisher; Fineness of Gold With Special Reference to the Morobe Gold Field, New Guinea: Econ. Geol., Vol. 40, No. 7, November, 1945.

HEAVY MINERALS FOUND IN
WESTERN GOLD PLACERS[1]

MINERAL	SPECIFIC GRAVITY	
	FROM	TO
Magnetite		5.2
Ilmenite	5.6	5.7
Garnet	3.2	4.3
Zircon	4.2	4.7
Hematite	4.9	5.3
Chromite	4.3	4.6
Epidote	3.2	3.5
Olivine		3.3
Limonite	3.6	4.0
Rutile		4.2
Pyroxine		3.3
Monazite	4.9	5.3
Platinum group metals	14	19

[1] A partial list, arranged in approximate order of frequency.

Gold, platinum, magetite, ilmenite, chromite, garnet, zircon, rutile and monazite, when found in placers, may be far from their source. Copper minerals, galena and sphalerite are rarely found in placers, and where found, are usually not far from their source.

The black sand content of gold placers in the Western United States is commonly between 5 and 20 pounds of black sand concentrate per cubic yard of gravel.

APPROXIMATE MINING COSTS, 1964 [1]

Bucket-line dredge	11c - 22c per cubic yard
Dragline dredge	22c - 45c " " "
Hydraulic	25c - 45c " " "
Drift	$9.00 - $18.00 " "

BUCKET-LINE DREDGE COST, 1964 [2]

Annual capacity, million cubic yards	Bucket size, cubic feet	Maximum digging depth below water level, feet	Weight, tons	Cost per ton, erected, dollars	Cost, erected, dollars (million)
8.4.	22	150*	5,000	1,300	6.5
10.4.	27	120	4,000	1,370	5.5
6.9.	18	100	3,000	1,400	4.2
3.8.	10	80	2,000	1,600	3.2
3.0.	8	50	1,300	1,850	2.4
2.3.	6	40	700	2,000	1.4
1.6.	4¼	25	460	2,175	1.0

* A bucket-line dredge of this digging capacity has not been built; bucket size may be restricted by engineering requirements.

FLOATING GRAVEL WASHING PLANT COST, 1964 [2] [3]

Annual capacity, million cubic yards	Trommel size	Weight short tons	Cost per ton, erected, dollars	Cost, erected, thousand dollars
0.6.	54-inch diameter, by 30 feet.	76	1,380	105
1.5.	72-inch diameter, by 47 feet.	207	1,210	248
3.0.	120-inch diameter, by 63 feet.	488	1,270	620

1/ U.S. Bureau of Mines I.C. 8331, p. 13, 1967.
2/ U.S. Bureau of Mines I.C. 8331, p. 14, 1967.
3/ Washing plant for dragline dredge.

APPENDIX F

CONVERSION TABLES

Lengths

1 mile =
- 8 furlongs
- 80 chains
- 320 rods
- 1,760 yards
- 5,280 feet

1 furlong =
- 10 chains
- 220 yards

1 station =
- 6.06 rods
- 33.3 yards
- 100 feet

1 chain =
- 4 rods
- 22 yards
- 66 feet
- 100 links

1 rod =
- 5.5 yards
- 16.5 feet

1 yard =
- 3 feet
- 36 inches

1 vara = 33 inches (approx.)
1 foot = 12 inches
1 link = 7.92 inches
1 inch = 0.0833 foot

Square Measure

1 township = 36 sq. miles

1 sq. mile =
- 1 section
- 640 acres

1 acre =
- 4,840 sq. yards
- 43,560 sq. feet
- 10 sq. chains
- 160 sq. rods

Lode Claim =
- 600 ft. x 1,500 ft.
- 20.661 acres
- 3305.78 sq. rods

Placer Claim = 20 acres
(1 locator)
1 sq. rod = 272¼ sq. feet
1 sq. yard = 9 sq. feet
1 sq. foot = 144 sq. inches

Cubic Measure

1 cubic yard = 27 cubic feet
1 cord (wood) = 4 x 4 x 8 ft. =
    128 cu. ft.
1 ton (shipping) = 40 cubic ft.
1 cubic foot = 1,728 cubic inches

1 cu. ft. = 7.48 U.S. gallons
1 bushel = 2150.42 cu. in.
1 bushel = 1.244 + cu. ft.
1 gallon = 231 cu. in.

Weights (Commercial)

1 long ton = 2,240 lbs.
1 short ton = 2,000 lbs.

1 pound = 16 ounces
1 ounce = 16 drams

Troy Weight (for Gold and Silver)

12 oz. troy = 1 lb. troy = 0.823 pounds av. = 5760 grains
16 oz. av. = 1 lb. av. = 7,000 grains
31.1035 grams = 1 oz. troy = 20 pennyweight = 480 grains
28.350 grams = 1 oz. av.
1 kilogram = 2.2046 lb. av.
1 pennyweight = 24 grains = 1.555 grams

Dry Measure

1 bushel =
- 4 pecks
- 32 quarts

1 peck = 8 quarts

1 quart = 2 pints
1 bushel = 1.2445 cubic feet

Metric and U.S. Weights and Measures

Lengths

Miles×1.6093 = Kilometers
Yards×.9144 = Meters
Feet×.3048 = Meters
Feet×30.48 = Centimeters
Inches×2.54 = Centimeters
Inches×25.4 = Millimeters
Kilometers×.621 = Miles

Kilometers×1093.6 = Yards
Kilometers×3280.9 = Feet
Meters×1.094 = Yards
Meters×3.281 = Feet
Meters×39.37 = Inches
Centimeters×.3937 = Inches
Millimeters×.03937 = Inches

Areas

Sq. mile×2.59 = Sq. kilometers
Acres×.00405 = Sq. kilometers
Acres×.4047 = Hectares
Sq. Yards×.8361 = Sq. meters
Sq. feet×.0929 = Sq. meters
Sq. Inches×6.452 = Sq. centimeters
Sq. inches×645.2 = Sq. millimeters
Sq. kilometers×.3861 = Sq. miles
Sq. kilometers×247.11 = Acres
Hectares×2.471 = Acres
Sq. meters×1.196 = Sq. yards
Sq. meters×10.764 = Sq. feet
Sq. centimeters×.155 = Sq. inches
Sq. millimeters×.00155 = Sq. inches

Volume

Cu. yards×.765 = Cu. meters
Cu. feet×.0283 = Cu. meters
Cu. inches×16.383 = Cu. centimeters
Cu. meters×1.308 = Cu. yards
Cu. meters×35.3145 = Cu. feet
Cu. centimeters×.06102 = Cu. inches

Liquid Measure

U.S. gallons×.8333 = Imperial gallons
Gallons×3.785 = Litres
Quarts×.946 = Litres
Imperial gallons×1.2009 = U.S. gallons
Litres×.2642 = Gallons
Litres×1.057 = Quarts

Weights

Pounds×.453 = Kilograms

Kilograms×2.2046 = Pounds

INDEX

Caliche, as factor in dry washing, 83
  cemented gravels, 41
  described, 123
  residual-type enrichments on 17
California, auriferous conglomerate, in, 5
  beach placers, 20, 21
  El Dorado County, seam diggings 13
  Tertiary gravels in, 4, 18, 19, 61
Canvas, use in rocker, 79, 80
  use in sampling sluice, 75
Casing, churn drill, 123
  drilling with, 44-46, 171-175
  drilling without, 49, 64-65
Casing factor, 123
Cement, 123, 124
Channel cut, use in sampling, 37-38, 62
  size of, 34, 37
Channel obstructions, effect on gold
  deposition, 6
Channel scour, as factor in gold
  deposition, 7
Character of placer gold, related to
  distance from source, 192
Check holes, interpreting, 55
Check list, for placer investigations, 107-115
Check samples, as "written proof"
  where no values exist, 60
  interpreting, 55
  to detect salting, 67
  use in valuation, 28
Check shafts, use in placer drilling,
  36, 55, 60
Chellson, Harry C., cited 101
Churn drill, 124
  multiplying factors for, 184-186
  use in drift mine prospecting, 61
  use in hydraulic mine prospecting, 62
  use in sampling, 44-57
Churn drill, as a method for testing
  placers, 44-57
  computing values from, 46-49, 184-186
  cost of, 55-57
  equipment for, 44-46, 171-177
  excess or deficient core, causes of, 47
  frozen ground, 49
  hole spacing, 51-55
  holes per acre, 33, 52
  pattern, for narrow deposits, 51-53

  pattern, for widespread deposits, 52-54
  personnel for, 50-51
  procedure, 44-46, 171-177
  rate of drilling, 55-57
  ratio, holes per acre, 52, 53
  ratio, sample to gravel, 33
  supervision of, 50-51
  value adjustments, 47-48, 186
  without casing, 49, 64
Choosing a sampling method, 30, 31

Clamshell-type excavator, use in
  sampling, 38, 40-41
Clark, William B., cited, 13, 18
Clean-up, definition of, 124
Clifton, H. Edward, cited, 21
Climatic change, effect on placers, 7
Coarse gold, 124
  effect on samples, 28, 57, 98, 99
Cocoa matting, 124
  use in sampling sluice, 75
Colloidal gold, 124
  in unproven processes, 103
Colluvial definition of, 124
Color, gold definition of, 124
  standard sizes of, 93
Colorado, gold from glacial moraines,
  5, 22
Concentration, dry, principles of, 82-83
  of gold on bedrock, 6-8
  of valuable minerals, 7-9
Concentrates, extracting gold from, 88-89,
  92, 93
Core, churn drill, 125
  factor, 125
  rise, 125
Corrected gold weight, in drill samples,
  47-49, 186
Conversion tables, water, 166-167
  weights and measures, 197-198
Correction factors, applied to samples,
  28, 60, 63, 83
  for churn drill, 186
Conversion tables, weights and measures
  197-198
Cost, factors influencing, 100
  of bucket-line dredges, 193
  of churn drilling, 55-57
  of sampling, with backhoe excavator,
  43
  of washing plants, 193
Cost data, sources of, 100-102
Cost estimates, making, 100-101
  based on hypothetical
    operation, 101
  comparison method, 100-101
Cost index, Engineering News-Record, 101
Costs, related to throughput, 103
  mining, 193
Cradle, (see rocker)
Creek placers, 13-14, 125
Crevicing, 125
Cribbing, 125
  diagram showing, 38
  use of, 38
Cutting samples, in pits or shafts, 37
Daily, Arthur, cited, 48, 53
Dasher, John, cited, 21
Data summary sheet, for samples, 155
Degrading stream, 6

Muck, Alaskan, description of, 137
Mud box (see puddling box)
Muddy water, consideration of, 107
   in dredge ponds, 166
Narrow pay channel, drilling pattern
   for, 51-52
   value calculation of, 54,55, 58-59
Nevada, Goldfield District, absence of
   placers in, 4
Nome, Alaska, beach placers, 9, 20
   elevated beaches, 9
Nugget, definition of, 137
Nuggety ground, use of bulk samples
   for, 57
Number of samples needed, 28,
   31-34, 52
Operation of churn drill, 172-175
Ore reserves, calculation of, 52-59
Oregon, beach placers, 20, 21
   Tertiary gravels, 18
Outwash aprons, placers in, 22
Over-valuation, from fire assays, 91
   from rotary drills, 65
Pan, 138
   amalgamation in, 92
   bluing before use, 89
   description of, 138
   factor, 89, 90. 138
   grizzly, use of, 89-90
   perforated, use of, 89-90
   removing grease from 89
   safety, use of, 90
   use as geologic tool, 90-91
   use for washing sample, 74
Panner, for placer drill, 46, 50-51
Panning, procedure described, 87-91
   notes and suggestions to improve, 89-91
   testing by, 34-35, 36, 61, 62
PAR-X mechanical shaft digger, 40
Pardee, J. T., cited, 21
Patent, 138
Particle size, as function of stream
   processes, 7
Particle sizes, to be delt with in
   sampling, 27
Parting solution, preparation of, 93
Patman, Charles G., cited, 102
Pay lead, 138
   in drift mines, 60-61
Pay streak, 8-9, 17, 28, 138, 139
Pediment, 139
Penneyweight, 139
Perforated pan, use of, 89-90
Permafrost, 139
Peterson, Donald W., cited, 19
Pilot plant, use in sampling, 57
Pilot sluice, use on gold dredge, 139
Pinched sluice, 139
Pipe clay, 139

Pipe factor, 139
Placer, definition of, 140
   deposits, classification of, 13
   drill, basic equipment, illustration
      of, 45
   drill, typical set up, illustration
      of, 52
   drills, types of, 49
   geology, review of, 3-9
   machines, small scale, 81-82
   material, typical, photo of, 27
   mining, definition of, 140
   terms, glossary of, 119-149
   theory, review of, 3-9
Placers, accretion, 16, 119
   beach, 20-21, 143
   bench, 15-16, 121
   burial, preservation by, 9
   buried, 123
   classification of, 13
   creek, 14, 125
   defined, 3, 140
   desert, 17-18
   dry, 128
   eluvial, 13-14
   eolian, 22
   flood gold, 16, 17, 130
   glacial, 5, 21, 22
   gravel-plain, 15, 132
   gulch, 14
   marine, 136
   residual, 13, 141
   resorted, 5, 22
   river, 14-15
   skim bar, 16, 143-144
   stream, 14-15
   study of, 3-9
   Tertiary, 18-19, 146
Plant run, use in sampling, 57
Platinum placers, 5
Platinum-group minerals, in beach
   placers, 20-21
Point bar, 143, 144
Pre-existing placers, as source of gold, 4
Presampling reconnaissance, 30
Preservation of placers, 9
Processing record, for samples, 154
Production records, use of, 30, 62, 68-69
Prommel, H. W. C., cited, 40
Prospecting, definition of, 140
Prospecting rockers, construction
   details, 77-79
Puddling box, use with sampling
   sluice, 76
Quaternary gravels, 140
Quicksilver (mercury), 136
   effect on surface texture of gold, 75, 93
   use in pan, 88
   use in riffles, 75

Terrace—Continued
   gravels, 9
Tertiary gravels, 9, 13, 18-19, 61
Tertiary period, geologic chart
   showing, 146
Tertiary River systems, 4, 147
Theobald, Paul K., Jr., cited, 91
Thomas, Bruce I., cited, 21, 102
Thompson, Arthur G., cited, 21
Thurman, Charles, H., cited, 102
Trace, gold definition of, 147
Trained personnel, need in placer
   drilling, 48, 50-51
Trestle sluice, 147
Triangle method, for calculating
   reserves, 52-54, 56
Trommel, 147
Troy ounce, definition of, 147
Troy weight, conversion table for, 160
Twenhofel, W. H., cited, 21
Types of placers, 13-24
Uncased drill holes, sampling
   problems, 64-66
Undercurrent, 147
Unproven processes, 103
Ultrabasic rocks, as platinum sources, 5
Valuable minerals, sources of, 4-5
Valuation, area of influence in, 32
   factors, other than mineral content,
      29, 107
   initial, 33-34
   of placer deposits, calculating
      procedures 52-58
Value of deposit, calculated by mean-area
   method, 54-55,58-59
   calculated by trangle method, 52-54, 56
Values, definition of, 148
   reporting, 99
Vanderburg, William O., cited 14, 18, 93
Volume, error in, effect on sample value,
   65-66
   factor, churn drill, 148

Volume—Continued
   measurements, in sampling, 36
Ward-type placer drill, 50, 148
Wash, definitions of, 148
Washing equipment, for samples, 73-83
Washing plants, cost of, 193
   water requirements for, 165-166
Water duty, 128
Water head, 132
   measurements, units of, 166-167
Water requirements for, bucket-line
   dredges, 166
   dragline dredges, 166
   dryland dredges, 166
   ground sluicing, 165
   hand mining, 165
   hydraulicking, 165
   moveable washing plants, 166
   placer operations, 165-166
   rockers 80, 165
   sampling sluice 76
   stationary washing plants, 165
Weathering, 148
Weathering and release processes 5-6
Weaverville, California, gold from
   glacial till, 5
Weight of sample, as volume check, 37
Weights of materials, 191
Wet ground, cribbing for illustrated,38
   use of caissons in, 37
Whitney, J. D., cited 18-19
   quoted, 61
Widespread placers, prospecting by use
   of rectangular grid, 52-54, 56
Wilson, Eldred D., cited, 18
   quoted, 83
Wind-caused surface enrichment, 4,
   17-18, 22
Woodbridge, T. R., cited, 32
Wyoming, Snake River, flood gold in, 17
Yardage, definitions of, 148
Zircon, 149

⌃ U. S. GOVERNMENT PRINTING OFFICE : 1969 O - 353-917